BIOENERGY FROM AGRICULTURAL RESIDUES

Woodhead Series in Bioenergy

The *Woodhead Series in Bioenergy* publishes titles on the topic of bioenergy as it relates to renewable sources of energy. The series notably covers biological resources, chemical and biological processes, and biomass products for energy production, along with the environmental, social, management, and economic aspects of bioenergy.

Series Editor

Prof. Yves Gagnon, Université de Moncton, Moncton, Canada

Previous Volumes in the Series

Kishor Kumar Sadasivuni and Mithra Geetha, Biofuel Cells and Energy Generation, 2025, 9780443216022

- Krishna Kumar Jaiswal, Bhaskar Singh and Amit K. Jaiswal, Microalgal Biofuels: Sustainable Production, Conversion, and Life Cycle Assessment, 2025, 9780443241109
- Jeyabalan Sangeetha and Devarajan Thangadurai, Microalgal Biomass for Bioenergy Applications, 2024, 9780443139277
- Lina María Grajales Agudelo, Juan Carlos Valdés Serra and Erich Collicchio, Agroenergy, 2024, 9780443214301
- Stefano Ferrari Interlenghi, José Luiz de Medeiros and Ofélia de Queiroz Fernandes Araújo, Sustainability of Methylic and Ethylic Biodiesel Production Routes, 2024, 9780443219429
- Hwai Chyuan Ong, Islam Md Rizwanul Fattah and Indra Mahlia, Waste Valorization for Bioenergy and Bioproducts, 2024, 9780443191718

WOODHEAD SERIES IN BIOENERGY

BIOENERGY FROM AGRICULTURAL RESIDUES

Edited by

R. PRAVEEN KUMAR

Department of Biotechnology, Arunai Engineering College, Chennai; Tiruvannamalai, Tamil Nadu, India

B. BHARATHIRAJA

Department of Biotechnology, Easwari Engineering College, Chennai, Tamil Nadu, India

ELSEVIER

WP WOODHEAD PUBLISHING

An imprint of Elsevier

Woodhead Publishing is an imprint of Elsevier
50 Hampshire Street, 5th Floor, Cambridge, MA 02139, United States
125 London Wall, London EC2Y 5AS, United Kingdom

For accessibility purposes, images in electronic versions of this book are accompanied by alt text descriptions independently provided by Elsevier, generated with the aid of AI. For more information, see https://www.elsevier.com/about/accessibility.

Books and Journals published by Elsevier comply with applicable product safety requirements. For any product safety concerns or queries, please contact our authorised representative, Elsevier B.V., at productsafety@elsevier.com.

Publisher's note: Elsevier takes a neutral position with respect to territorial disputes or jurisdictional claims in its published content, including in maps and institutional affiliations.

Notices
Knowledge and best practice in this field are constantly changing. As new research and experience broaden our understanding, changes in research methods, professional practices, or medical treatment may become necessary.

Practitioners and researchers must always rely on their own experience and knowledge in evaluating and using any information, methods, compounds, or experiments described herein. In using such information or methods they should be mindful of their own safety and the safety of others, including parties for whom they have a professional responsibility.

To the fullest extent of the law, neither the Publisher nor the authors, contributors, or editors, assume any liability for any injury and/or damage to persons or property as a matter of products liability, negligence or otherwise, or from any use or operation of any methods, products, instructions, or ideas contained in the material herein.

ISBN: 978-0-443-24726-2 (print)
ISBN: 978-0-443-24727-9 (online)

For information on all Woodhead publications visit our website at https://www.elsevier.com/books-and-journals

Publisher: Priscilla Braglia
Acquisitions Editor: Ruchi Bhardwaj
Editorial Project Manager: Namrata Aggarwal
Production Project Manager: Anitha Sivaraj
Cover Designer: Gopalakrishnan Venkatraman

Typeset by Integra Software Services

Contents

Contributors

Gurunathan Baskar
Department of Biotechnology, St. Joseph's College of Engineering, Chennai, Tamil Nadu, India

B. Bharathiraja
Department of Biotechnology, Easwari Engineering College, Chennai, Tamil Nadu, India

Rahul Krishna Bhuvanendran
Department of Food Technology, TKM Institute of Technology, Kollam, Kerala, India

Matta Chenna Keshava Charan
Department of Biotechnology, Chaitanya Bharathi Institute of Technology, Hyderabad, Telangana, India

Balaji Doolam
Department of Biotechnology, Chaitanya Bharathi Institute of Technology, Hyderabad, Telangana, India

B.S. Harish
Department of Biotechnology, Easwari Engineering College, Chennai, Tamil Nadu, India

J. Jayamuthunagai
Centre for Biotechnology, Alagappa College of Technology, Anna University, Chennai, Tamil Nadu, India

Aravind Madhavan
School of Biotechnology, Amrita Vishwa Vidyapeetham, Amritapuri, Kollam, Kerala, India

Sanjeeb Kumar Mandal
Department of Biotechnology, Chaitanya Bharathi Institute of Technology, Hyderabad, Telangana, India

Bishwambhar Mishra
Department of Biotechnology, Chaitanya Bharathi Institute of Technology, Hyderabad, Telangana, India

A. Mohan
Department of Civil Engineering, Easwari Engineering College, Ramapuram, Chennai, Tamil Nadu, India

R. Praveen kumar
Department of Biotechnology, Arunai Engineering College, Chennai; Tiruvannamalai, Tamil Nadu, India

Ravichandran Pravin
Department of Biotechnology, St. Joseph's College of Engineering, Chennai, Tamil Nadu, India

Reshmy Rajasekharan
Cellupro Green Pvt. Ltd., Industrial Development Area, Mavelikara, Kerala, India

Lakshmipriya Rajendran
Department of Food Technology, TKM Institute of Technology, Kollam, Kerala, India

Navnit Kumar Ramamoorthy
Department of Biotechnology, School of Lifesciences, Central University of Rajasthan, Ajmer, Rajasthan, India

C. Nagendranatha Reddy
Department of Biotechnology, Chaitanya Bharathi Institute of Technology, Hyderabad, Telangana, India

Samuel Lalthazuala Rokhum
Department of Chemistry, National Institute of Technology Silchar, Silchar, Assam, India

Sumithra Salla
Department of Biotechnology, Chaitanya Bharathi Institute of Technology, Hyderabad, Telangana, India

Kotteeswaran Santhanam
Department of Civil Engineering, SRM Institute of Science and Technology, Ramapuram, Chennai, Tamil Nadu, India

I. Abernaebenezer Selvakumari
Centre for Biotechnology, Alagappa College of Technology, Anna University, Chennai, Tamil Nadu, India

Kiruthika Shanmugam
Centre for Biotechnology, Alagappa College of Technology, Anna University, Chennai, Tamil Nadu, India

Raveendran Sindhu
Department of Food Technology, TKM Institute of Technology, Kollam; Santhigiri Scientific and Industrial Research Institute (SSIRI), Trivandrum, Kerala, India

Ninu Sunil
Department of Food Technology, TKM Institute of Technology, Kollam, Kerala, India

K. Thavasilingam
Department of Mechanical Engineering, Easwari Engineering College, Ramapuram, Chennai, Tamil Nadu, India

Deepa Thomas
Post Graduate and Research Department of Chemistry, Bishop Moore College, Mavelikara, Kerala, India

Rekha Unni
Post Graduate Department of Chemistry, Christian College, Chengannur, Kerala, India

S. Prasanna Raj Yadav
Department of Mechanical Engineering, Easwari Engineering College, Ramapuram, Chennai, Tamil Nadu, India

Rajasri Yadavalli
Department of Biotechnology, Chaitanya Bharathi Institute of Technology, Hyderabad, Telangana, India

D. Gayathri
Department of Computer Science and Engineering, Velammal Engineering College, Surapet, Chennai, India

Foreword

The energy sector is undergoing a significant transition to meet growing energy demands, reduce greenhouse gas emissions, support climate change mitigation, and protect the planet for future generations. A major component of this energy transition is the integration of large-scale renewable energy sources, including bioenergy, into the overall energy mix of countries.

Bioenergy refers to the conversion of biomass, derived from recently living organic materials, into electricity, heat, and transportation fuels. Alongside wind and solar energy, bioenergy is poised to play a significant role in the global energy transition.

To support both the research community and the commercial bioenergy sector, the Woodhead Series in Bioenergy publishes research-driven and application-focused titles that explore bioenergy as a renewable source of energy. The series covers biological resources, chemical and biological processes, and biomass-based products for energy production, all framed within the principles of sustainability, encompassing environmental, social, managerial, and economic dimensions.

Agricultural biomass feedstocks in the form of crop residues, animal waste and forestry residues are increasingly gaining attention as a renewable and sustainable source of energy.

Building on their extensive research in the field of bioenergy, Prof. R. Praveen Kumar, from Arunai Engineering College, and Prof. B. Bharathiraja, from Easwari Engineering College, both from India, have led the development of this book focusing on the value chain of agricultural residues for bioenergy production.

Co-authored by the volume editors and other researchers, *Bioenergy from Agricultural Residues* provides an in-depth understanding of the latest developments in the field of bioenergy production from agricultural residues, the associated challenges, and the opportunities for future research and development. The book also details case studies and examples from successful bioenergy projects, providing insights into the establishment of sustainable bioenergy systems.

After an initial chapter introducing bioenergy and agricultural residues, a specific chapter details agricultural residues as a feedstock for bioenergy production. This is followed by four chapters on the various technological conversion processes to transform agricultural residues into bioenergy (biogas, bioethanol, biodiesel, pyrolysis and gasification), and two chapters on the sustainability issues (life cycle assessment, economics, and policies) of bioenergy production from agricultural residues. Finally, the last chapter presents future prospects and challenges for bioenergy production from agricultural residues.

This title will be of interest to academic and industrial researchers, agriculture and industry practitioners, engineers, scientists, along with graduate and post-graduate students working in the valorisation of agricultural residues for bioenergy production.

Yves Gagnon P.Eng. D.Sc.
Université de Moncton, Canada
Series Editor, Woodhead Series in Bioenergy

CHAPTER 1

Introduction to bioenergy and agricultural residues

B.S. Harish[a], R. Praveen Kumar[b], and B. Bharathiraja[a]
[a]Department of Biotechnology, Easwari Engineering College, Chennai, Tamil Nadu, India
[b]Department of Biotechnology, Arunai Engineering College, Chennai, Tamil Nadu, India

1.1 Definition and importance of bioenergy

Bioenergy is the renewable energy produced by converting organic biomass into sustainable fuels through biological, thermochemical, and/or physicochemical routes. Bioenergy has gained significant attention in recent years due to its carbon-neutral nature and its potential to mitigate the adverse environmental impacts associated with extensive reliance on nonrenewable energy sources. Bioenergy, produced by the oxidation or conversion of biomass substrates, is broadly classified into first-generation (edible food crops), second-generation (nonedible lignocellulosic materials), and third-generation (algal biomass) (Chowdhury et al., 2025).

Biomass conversion routes into bioenergy depend on the type of feedstock and the energy product. Biological routes employ microbial and enzymatic processes such as anaerobic digestion (AD) and fermentation for the production of biofuels. These methods are particularly effective for wet, biodegradable residues, though pretreatment is often needed to improve digestibility and yields. Thermochemical routes are typically applied to dry biomass and include combustion for direct heat and power generation; gasification and pyrolysis for producing syngas, bio-oil, and char; as well as hydrothermal liquefaction for liquid fuel precursors. Physicochemical or chemical conversion processes, most notably transesterification, are primarily used for lipid-rich feedstocks to obtain biodiesel, with catalytic upgrading and refining techniques enhancing fuel quality and energy density. Increasingly, hybrid and integrated conversion systems are being developed to handle diverse feedstocks and improve overall energy recovery; however, large-scale deployment still faces challenges such as feedstock variability, longer processing times, and the need for more sustainable and cost-effective pretreatment technologies (Saravanan et al., 2025).

Bioenergy from Agricultural Residues
https://doi.org/10.1016/B978-0-443-24726-2.00003-1

At the global level, bioenergy is projected to play a pivotal role in decarbonization strategies and efforts to achieve net-zero greenhouse gas (GHG) emissions (Reid et al., 2020). Large-scale bioenergy production is therefore considered crucial for climate change mitigation and future energy supply scenarios. According to the International Energy Agency, biomass could supply an additional 50 EJ (approximately 10%) of global primary energy by 2035, with an even higher potential if advances in technology and sustainability frameworks are realized. Beyond its climate mitigation potential, bioenergy contributes to energy security by diversifying the renewable energy mix and can stimulate rural economic development through new markets for agricultural residues and other biomass resources (Slade et al., 2014).

In essence, bioenergy represents a versatile and strategically significant renewable energy pathway that not only diversifies the global energy mix but also contributes to reducing GHG emissions, enhancing rural livelihoods, and promoting sustainable resource management (Asaad et al., 2023).

1.2 Overview of agricultural residues

Agricultural residues represent one of the most abundant yet underutilized biomass resources worldwide, with global production estimated at nearly **1300 million tons annually**. Major contributors include **China (91.6 million tons), India (68.8 million tons), Nigeria (37.9 million tons), and the United States (19.4 million tons)** of food-related agricultural waste each year (Egwuatu et al., 2024).

Agricultural residues can be broadly classified into two main categories (Honorato-Salazar and Sadhukhan, 2020):

- **Field-based (primary) residues**—Left in the field after crop harvesting, including **cereal straws (wheat, rice, barley, and maize stover)**, **sugarcane trash**, stalks, leaves, and shells.
- **Processing-based (secondary) residues**—Agro-industrial by-products generated during postharvest industrial processing, such as rice husks, sugarcane bagasse, coconut shells, oilseed cakes, corn cobs, and residues from fruit and vegetable processing. Fig. 1.1

Among these, crop residues are particularly significant due to their high **lignocellulosic content**, primarily composed of **cellulose, hemicellulose, and lignin**, with minor fractions of extractives and ash. This composition provides abundant carbon resources for conversion into fuels, chemicals, and other materials (Shahzad et al., 2024). However, their conversion

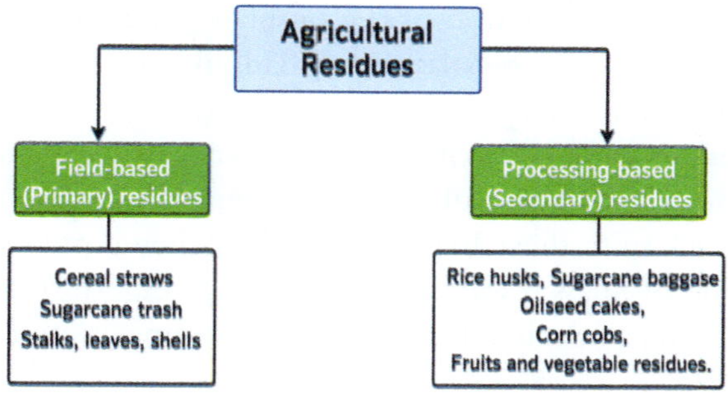

Fig. 1.1 Classification of agricultural residues.

efficiency is limited by the **recalcitrant nature of lignin and crystalline cellulose**, which restricts enzymatic hydrolysis. To overcome this, a range of **pretreatment techniques**—physical (milling), chemical (dilute acid or alkali), biological (microbes and fungi), and enzymatic hydrolysis—are employed to improve substrate accessibility.

Despite their vast availability and potential, agricultural residues are often **underutilized or mismanaged**. Traditional practices such as **open-field burning, landfilling, composting, and limited use as animal feed** remain dominant. In South and East Asia, large-scale burning of rice and wheat straw contributes to **air pollution, GHG emissions, and smog episodes**. Landfilling leads to **methane emissions** and land degradation, while composting and AD, although more sustainable, remain underutilized relative to the scale of residue generation.

Harnessing agricultural residues for bioenergy offers multiple co-benefits, including enhanced rural energy access and the promotion of circular bioeconomy strategies. With improvements in **collection logistics, sustainable residue management practices, and supportive policy frameworks**, these resources could significantly expand the renewable energy supply while maintaining **soil health and agricultural productivity** (Caicedo et al., 2016).

1.3 Field-based (primary) crop residues

Field-based agricultural residues, often referred to as primary residues, are the portions of crops left behind in the field after harvesting. These include straw, stalks, stems, leaves, husks, and cobs, which together constitute a significant fraction of the biomass generated in global agricultural systems (Honorato-Salazar and Sadhukhan, 2020). Unlike processing-based residues that accumulate near agro-industrial units, field-based residues are widely distributed across landscapes, often scattered in low-density forms, and are subject to challenges in collection, storage, and transport. Despite these logistical barriers, they represent one of the most promising biomass resources for bioenergy due to their abundance, renewable nature, and rich lignocellulosic composition (Lackner and Besharati, 2025).

A key advantage of field-based residues is their high lignocellulosic content, which is suitable for multiple bioenergy conversion pathways. For instance, cereal straws such as wheat, rice, and barley straw, along with maize stover and sugarcane trash, are rich in fermentable polysaccharides that can be hydrolyzed and converted into bioethanol through biochemical routes. Similarly, their fibrous structure makes them amenable to thermochemical processes such as pyrolysis, gasification, and combustion for producing heat, electricity, and syngas (Shahzad et al., 2024; Pathak and Kothari, 2022). This versatility highlights their role as a critical component in the development of sustainable bioenergy systems.

However, the use of field-based residues is constrained by competing demands. These residues are often incorporated into soils to maintain organic matter, enhance fertility, and reduce erosion. Removing excessive amounts can negatively impact soil health, nutrient cycling, and long-term agricultural productivity (Aleminew, 2024). Hence, sustainable removal rates must be carefully determined, balancing energy recovery with the preservation of ecosystem services. Studies emphasize that partial removal of residues, combined with conservation practices, can create a balance between energy use and agricultural sustainability (Kemausuor et al., 2014; Blanco-Canqui and Lal, 2009).

From a technological standpoint, field residues pose challenges due to their heterogeneous physical and chemical properties. High ash and silica content, especially in rice straw, leads to slagging and fouling during combustion (Dizaji et al., 2019). Similarly, the high moisture content of freshly harvested residues reduces their energy density and increases transport costs. Pretreatment strategies such as drying, torrefaction, or pelletization can improve handling characteristics and conversion efficiencies (Jenkins and Sumner, 1986). Furthermore, advances in biochemical conversion—such as enzyme engineering for efficient cellulose hydrolysis—are expanding the feasibility of utilizing recalcitrant lignocellulosic residues for liquid biofuels.

1.4 Processing-based (secondary) crop residues

Processing-based residues are generated during postharvest and agro-industrial operations, and they represent a concentrated and readily available biomass stream for bioenergy applications. Typical examples include rice husk from milling, sugarcane bagasse from juice extraction, oilseed cakes from pressing, fruit pomace from juice and wine production, coconut shells, and nut shells (Honorato-Salazar and Sadhukhan, 2020). These residues are often accumulated in large volumes near processing units, making collection and transport more feasible compared to field-based residues. Their composition varies widely: lignocellulosic residues such as husks, shells, and bagasse are rich in cellulose, hemicellulose, and lignin, making them suitable for thermochemical and biochemical conversion, while oilseed cakes and fruit pomace contain significant amounts of proteins, starches, and lipids that can be harnessed for biodiesel or biogas production (Kumar et al., 2008). However, challenges such as high moisture content, seasonal availability, and potential contamination need to be addressed for efficient utilization. When effectively valorized, processing-based residues not only reduce agro-industrial waste management burdens but also contribute substantially to renewable energy generation, circular bioeconomy development, and rural industrial sustainability (Tables 1.1 and 1.2).

1.5 Current state of bioenergy from agricultural residues

Agricultural residues are increasingly recognized as one of the most promising feedstocks for bioenergy production due to their abundance, renewability, and minimal conflict with food resources (Sarangi and Nayak, 2021). Globally, crop residues such as rice straw, wheat straw, maize stover, sugarcane bagasse, and oilseed cakes form the backbone of second-generation biofuel research and development. Despite their vast availability, only a small fraction is currently utilized for bioenergy, with most residues either left in the field, burned, or used in low-value applications. Recent advancements in biochemical, thermochemical, and hybrid technologies have stimulated interest in their large-scale deployment. The following

Table 1.1 Primary agricultural residues as feedstock for bioenergy production.

Feedstock	Residues	Process	Parameters	Bioenergy form	Yield product	Reference
Vineyard and olive groves (Umbria region)	Pruning residues (vine shoots, olive branches)	Combustion	–	Bioelectricity and thermal	–	Torquati et al. (2016)
Rice	Surplus rice straw after harvesting	Hydrolysis and fermentation	DA: dilute H_2SO_4, ~160°C, ~10 bar; SE: ~200–220°C, ~20–25 bar	Bioethanol	DA:5295 and SE: 6928 GWh (estimated)	Hassan et al. (2021)
Corn	Corn Stalk	Anaerobic fermentation with biochar from residue cornstalk left after anaerobic fermentation (RCA-biochar) supplementation	RCA-biochar concentration: 15 g/L, pH-7.5, 60°C, 150 rpm, 96 h, inoculum: *T. thermosaccharolyticum* M18 0.2 g/L	Biohydrogen	286.1 mL H_2/g substrate	Zhao et al. (2021)
Oat straw	Lignocellulosic straw	Pyrolysis	Temperature: 300–600°C; Atmosphere: Nitrogen; Reactor: semi-batch vertical; Residence time: 1–38 min;	Syngas, Bio-oil, Biochar	Gas yield ↑ with temperature (13.1% at 300°C → 16.3% at 600°C); Tar yield max 57% at 600°C; Char yield ↓ with temp (48% at 300°C → 24% at 600°C)	Mlonka-Mędrala et al. (2021)

Table 1.2 Secondary agricultural residues as feedstock for bioenergy production.

Feedstock	Residues	Process	Parameters	Bioenergy form	Yield product	Reference
Rice husk	Secondary	Pyrolysis	Rice hull ash as catalyst; Heating rates: 10, 20, 50, 100 K/min	Syngas, Bio-oil, Biochar	High Heating Value of 15.49 MJ/kg	Loy et al. (2018)
Sugarcane bagasse (SCB)	Secondary agro-industrial residue (sugar industry by-product)	**Simultaneous saccharification and fermentation (SSF)** using *Saccharomyces cerevisiae*	**Simultaneous saccharification and fermentation (SSF)** using *Saccharomyces cerevisiae*	Bioethanol	4.88 g/L	Yanchal Jugwanth and Kana (2020)
Jatropha curcas seeds	Seed cake waste (after oil extraction)	Oil extraction (hexane) + Transesterification (alkali/acid catalysis)	Methanol under alkaline conditions with NaOH as catalyst; 62°C for 1 h.	Biodiesel	95.5% transesterification	Sánchez-Arreola et al. (2015)
Corncob	Secondary residue	Thermochemical pretreatment followed by dark fermentation	Pretreatment: 2% H_2SO_4 at 160°C, bovine ruminal fluid (BRF) bioaugmented with *Clostridium acetobutylicum (1:2)*, 35°C, pH 5.5	Biohydrogen	134.7 g/L	Medina-Morales et al. (2021)

Continued

Table 1.2 Secondary agricultural residues as feedstock for bioenergy production—cont'd

Feedstock	Residues	Process	Parameters	Bioenergy form	Yield product	Reference
Apple pomace	Agro-industrial by-product (fruit processing waste)	**Fermentation for bioethanol and anaerobic digestion for biomethane**	Pretreatment by 1% H_2SO_4, 140°C, 30 min	Bioethanol and biomethane	1 MT of pomace produced 164 kg pectin, 99 L bioethanol, and 33.6 m^3 methane	Vaez et al. (2023)

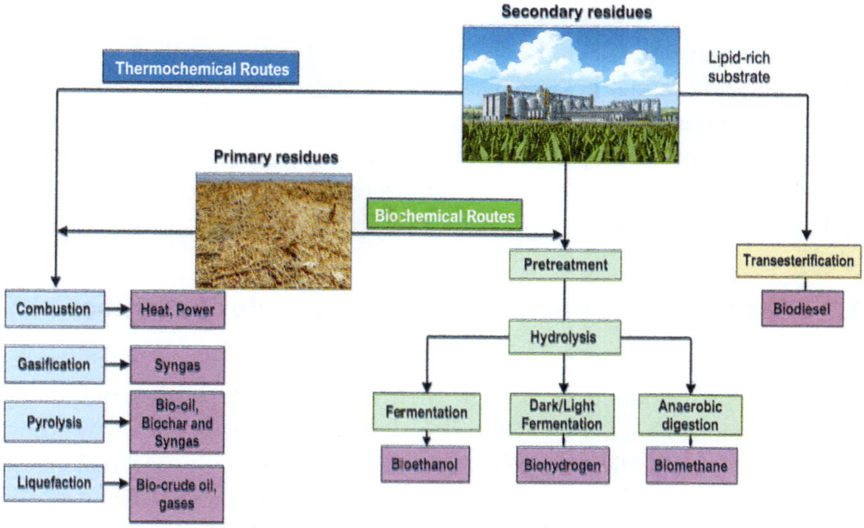

Fig. 1.2 Conversion pathways of agricultural residues into bioenergy through thermochemical, biochemical, and transesterification routes.

sub-sections outline the current state of bioenergy production pathways from agricultural residues (Fig. 1.2).

1.6 Thermochemical routes

Crop residues and agricultural wastes, classified as bio-renewable resources, have emerged as crucial feedstocks for bioenergy production. Thermochemical processes such as combustion, gasification, pyrolysis, and hydrothermal liquefaction convert these residues into heat, power, liquid fuels, and syngas, offering flexibility in accommodating diverse feedstocks and producing multiple value-added products (Muh et al., 2021). Combustion remains the most widely practiced method, accounting for >90% of renewable-based heating demand, particularly through the use of residues such as bagasse and rice husks in agro-industrial co-generation (Strzalka et al., 2017). Gasification produces syngas rich in CO and H_2, which can be upgraded for electricity and hydrogen production. Pyrolysis, by contrast, yields bio-oil, biochar, and light hydrocarbons, with product distribution highly dependent on feedstock type and processing conditions (You et al., 2018). Hydrothermal liquefaction provides the advantage of processing wet biomass without prior drying and has demonstrated high efficiencies with agricultural residues such as wheat straw and sugarcane bagasse (Ariyawansha et al., 2022). Despite their promise, challenges remain, including high capital costs, slagging and fouling due to ash content, and variability in feedstock properties. Nevertheless, thermochemical routes are a cornerstone of residue valorization and are increasingly being integrated with biological processes, such as syngas fermentation, to maximize fuel and chemical outputs.

1.7 Bioethanol production from lignocellulosic residues

Both primary and secondary agricultural residues represent vital feedstocks for bioethanol production, though their utilization differs in complexity. Primary residues, such as straws and stalks left in the field after harvesting, are relatively straightforward to assess in terms of availability and collection. In contrast, secondary residues arising from crop processing, such as husks, bagasse, and bran, present greater challenges. These challenges are linked to technical, economic, and environmental factors, including logistical constraints, variable feedstock quality, and competing market demands. Many of these residues already serve established purposes; for instance, corncobs are used in chemical manufacturing and energy generation, while rice husks and sugarcane bagasse are extensively employed in cogeneration plants. Redirecting such residues toward bioethanol production risks disrupting existing markets, driving competition, and elevating residue prices, thereby affecting the economic feasibility of bioethanol production (Same et al., 2024).

From a technological standpoint, lignocellulosic agricultural residues—including rice straw, wheat straw, corn stover, and sugarcane bagasse—are considered highly promising for bioethanol production due to their abundance of polysaccharides (cellulose and hemicellulose). Current conversion technologies typically involve a pretreatment stage (physical, chemical, or biological) to overcome the recalcitrance of lignin, followed by enzymatic hydrolysis to release fermentable sugars, which are then subjected to fermentation to produce ethanol. Despite advances, commercialization of cellulosic ethanol remains constrained by the high costs associated with pretreatment, the inefficiencies of hydrolytic enzymes, and difficulties in scaling fermentation systems to industrial levels.

Globally, significant efforts are underway to overcome these barriers. Countries such as Brazil, the United States, India, and China are investing heavily in pilot and demonstration plants, with sugarcane bagasse and corn stover serving as primary feedstocks. There is an increased focus in research on reducing costs through improved pretreatment strategies, development of robust enzyme systems, and the establishment of integrated biorefineries that valorize co-products such as lignin, which can be converted into energy or value-added chemicals (Sarangi and Nayak, 2021). These approaches aim to enhance process economics and drive the transition of cellulosic bioethanol from demonstration scale to commercial reality.

1.8 Biomethane production through AD

The expansion of biofuel production is increasingly linked to sustainable biomass sources, with agricultural residues recognized as one of the most critical feedstocks alongside forestry residues, municipal waste, and industrial by-products. Within this spectrum, both primary residues (e.g., rice straw, wheat straw, maize stover, and sugarcane trash left in fields after harvest) and secondary residues (e.g., rice husk, sugarcane bagasse, bran, fruit pomace, and oilseed cakes generated during agro-processing) provide abundant resources for biogas generation. Agricultural residues, largely lignocellulosic in nature, are highly resistant to microbial degradation during AD. The rigid lignocellulosic matrix restricts enzymatic accessibility, reducing methane yields. To overcome this challenge, pretreatment methods are commonly applied, including mechanical processing (milling and grinding), ultrasonic and thermal hydrolysis, and chemical, thermochemical, and biological treatments.

During the AD, the lignocellulosic material undergoes a series of microbial processes: hydrolysis of complex carbohydrates into simple sugars, acidogenesis to form volatile fatty acids (VFAs), acetogenesis, and finally methanogenesis, where methane and carbon dioxide are produced. The resulting biogas can be used directly for cooking, heating, or electricity generation or upgraded to biomethane for use as a transport fuel. A key advantage of this process is the production of digestate, a nutrient-rich by-product that retains much of the nitrogen, phosphorus, and other minerals present in the feedstock. Digestate serves as an effective soil amendment, improving fertility and crop yields while reducing dependence on synthetic fertilizers.

Globally, AD of agricultural residues is an established renewable energy pathway. In Europe, particularly in Germany and Denmark, large-scale biogas plants integrate crop residues, manure, and agro-industrial by-products into centralized energy systems. In Asia, small- to medium-scale digesters in India and China have played a critical role in rural electrification and provision of clean cooking fuel. Despite these successes, the use of lignocellulosic residues such as rice straw remains technically challenging due to their high lignin content and low biodegradability. To address this, co-digestion strategies—where crop residues are combined with more readily degradable substrates such as manure or food waste—are increasingly employed, improving nutrient balance and enhancing methane production.

Recent innovations are strengthening the feasibility of agricultural residue-based AD. Advances in microbial consortia engineering are enabling more efficient degradation of lignocellulose, while dry fermentation technologies are allowing higher solid loadings and reduced water consumption. In parallel, improvements in pretreatment techniques and biogas upgrading systems are making AD more economically viable and environmentally sustainable.

1.9 Biohydrogen production through dark and light fermentation

Biohydrogen has emerged as a promising clean fuel owing to its high energy content and zero carbon emissions during combustion. Among the various biological routes for hydrogen generation, dark fermentation (DF) and light-assisted processes such as photofermentation and biophotolysis are of particular significance. These methods not only facilitate renewable hydrogen production but also enable the valorization of organic wastes, creating dual benefits of energy recovery and waste minimization (Hitam and Jalil, 2023).

1.9.1 Dark fermentation

DF is one of the most established biological processes for hydrogen production. It is an anaerobic microbial pathway where carbohydrates are metabolized into hydrogen, carbon dioxide, and VFAs. Feedstocks such as agricultural residues, food-processing waste, paper industry effluents, and municipal organic waste are widely used as substrates. This process relies on hydrogenase- and nitrogenase-mediated enzymatic activities within fermentative microorganisms (Ntaikou et al., 2010).

The central metabolic pathway involves the conversion of glucose into pyruvate through glycolysis, generating ATP and NADH. Pyruvate can then be metabolized via different enzymatic routes depending on the microorganism and conditions. For instance, under strict anaerobic conditions, pyruvate–ferredoxin oxidoreductase converts pyruvate into acetyl-CoA and CO_2, while in facultative anaerobes, pyruvate formate lyase catalyzes the formation of acetyl-CoA and formate. The resulting acetyl-CoA is further converted into metabolites such as acetate, ethanol, or butyrate, each associated with hydrogen release. Among these pathways, the acetate route is generally more favorable for hydrogen generation (Albuquerque et al., 2024).

DF is particularly attractive because it is simple, relatively low-cost, and does not require light energy input. Moreover, it provides simultaneous benefits of sludge reduction and recovery of by-products such as short-chain fatty acids. However, its hydrogen yield is often constrained by competing microbial pathways such as methanogenesis and homoacetogenesis, which consume hydrogen. Strategies such as pH control, pretreatment of inoculum, and co-digestion with other substrates are often applied to improve yields.

1.9.2 Photo-fermentation

Photo-fermentation represents a light-dependent hydrogen production pathway driven by photosynthetic bacteria, including strains of Rhodobacter, Rhodospirillum, and Rhodopseudomonas. In this process, nitrogenase enzymes catalyze hydrogen production under illumination while decomposing organic compounds. A wide range of substrates can be utilized, including VFAs, sugars, and organic acids derived from agricultural and food wastes (Pandey et al., 2023).

The main advantage of photofermentation is its ability to use diverse feedstocks while generating high-purity hydrogen without oxygen contamination. The process operates at mild temperature and pressure conditions, making it environmentally benign. However, practical implementation faces challenges such as the low efficiency of light energy conversion, the need for sterile growth conditions, and the high energy demand of nitrogenase enzymes. Furthermore, reactor design plays a crucial role since light penetration is often hindered by cell density and shading effects. Despite these challenges, photofermentation is considered a promising complementary process to DF, especially when integrated in a two-stage system where DF-derived effluents rich in organic acids serve as substrates for photofermentation (Policastro et al., 2022).

1.10 Biodiesel production by transesterification of residues and by-products

Unlike bioethanol and biogas, biodiesel production from agricultural residues is less developed, as residues are typically lignocellulosic rather than lipid-rich. However, the potential of agricultural and agro-industrial sectors lies in their ability to generate large volumes of oil-rich residues. Waste vegetable oils and discarded cooking oils from households, food industries, and restaurants constitute a significant portion of this feedstock, alongside oilseed cakes and residues from crops such as soybean, cottonseed, sunflower, sesame, and rapeseed that remain after oil extraction. Animal by-products and fats derived from poultry houses, slaughterhouses, and fish-processing units also contribute valuable lipid-rich substrates, while microbial and algal oils, considered third-generation feedstocks, offer additional promise due to their high productivity and nonreliance on arable land. These feedstocks are especially attractive because they are relatively inexpensive, nonfood-based, and reduce dependence on edible oils, thereby minimizing competition with food security (Azadbakht et al., 2023).

The classification of biodiesel feedstocks has traditionally been organized into four generations. First-generation feedstocks are derived from edible oils such as soybean, corn, coconut, and palm oil, but their use is controversial due to food–fuel competition. Second-generation feedstocks consist of nonedible oils and oilseed by-products, including jatropha, neem, castor, and karanja, which are more sustainable since they grow on marginal lands and do not disrupt food systems. Third-generation feedstocks, comprising waste oils, animal fats, and algal oils, are increasingly attractive because they provide high lipid productivity and avoid competition with cropland. Finally, fourth-generation feedstocks include advanced biotechnological resources such as engineered microbial oils, photobiological solar fuels, and electro-fuels. From the perspective of agricultural residues and by-products, the most relevant categories are second- and third-generation feedstocks, as they directly exploit agro-residual streams without compromising food resources (Zambare et al., 2021).

The conversion of oils and fats derived from agricultural residues into biodiesel is achieved primarily through transesterification, a well-established process in which triglycerides react with an alcohol, typically methanol, in the presence of a catalyst to form fatty acid methyl esters and glycerol. This process remains the most widely adopted due to its high efficiency, scalability, and technological maturity. Nevertheless, alternative technologies such as pyrolysis, catalytic distillation, microemulsions, and microbial conversions are also under investigation, offering potential pathways to diversify biodiesel production. The efficiency and quality of biodiesel obtained are highly dependent on the characteristics of the feedstock, particularly free fatty acid levels, moisture content, and the presence of impurities. For this reason, pretreatment steps such as refining to remove phosphatides, pigments, and excess free fatty acids are often required to ensure both process efficiency and fuel quality (Wang et al., 2023).

1.11 Challenges and opportunities for the future

The use of crop residues for biofuel production has sparked ongoing debate about its implications for food and resource security. While second-generation biofuels rely largely on nonfood biomass such as straws, husks, and bagasse, large-scale diversion of these residues could still interfere with their traditional uses. Residues are commonly used as animal feed, soil amendments to maintain fertility, or raw material for industrial purposes. Redirecting them entirely toward bioenergy risks upsetting these established value chains and may indirectly affect food security and rural livelihoods. Careful management is therefore essential to balance the energy benefits with the broader needs of agricultural and local economies (Ali et al., 2019).

Another concern relates to soil health and biodiversity. Excessive removal of residues deprives soils of organic matter, which plays a critical role in maintaining structure, fertility, and microbial balance. The loss of organic carbon can accelerate erosion, nutrient depletion, and long-term productivity decline. At the same time, monocultures of biofuel crops pose risks to biodiversity and ecological balance. Sustainable harvesting practices, such as leaving a proportion of residues in the field, retaining nutrient-rich fractions, and adopting mixed cropping systems, are vital to preserve soil health and ecological stability while advancing bioenergy goals.

Water use and quality represent additional challenges. Crops such as maize, which are often cultivated for ethanol, require large amounts of irrigation and fertilizer (Liu et al., 2017). Their expansion for bioenergy purposes may exacerbate water scarcity and degrade water quality through nutrient runoff and eutrophication. Similarly, certain fast-growing bioenergy crops consume more water than traditional food crops or natural vegetation. To ensure sustainability, the deployment of residue-based biofuels must be paired with water-efficient cultivation practices and careful nutrient management to minimize environmental pressures.

Economic viability also constrains large-scale adoption. Biofuels often rely on subsidies, tax incentives, and favorable oil prices to remain competitive with fossil fuels. The high costs of residue collection, transportation, and processing make them less attractive in many regions. Furthermore, variability in feedstock availability and quality adds to economic uncertainty, complicating the design of stable supply chains. Addressing these challenges will require not only technological innovation but also supportive policies and infrastructure to reduce production costs (Cadenas and Cabezudo, 1998).

From a technological perspective, the conversion of lignocellulosic residues continues to face hurdles. Pretreatment methods are often energy-intensive and costly, while hydrolytic enzymes remain inefficient and expensive at scale. Fermentation systems and thermochemical processes such as gasification also struggle with technical limitations, particularly when dealing with residues that have high ash or moisture contents. These factors reduce conversion efficiency and raise operational costs. Advances in robust pretreatment technologies, cost-effective catalysts, and hybrid thermochemical–biological approaches are essential to unlock the full potential of residues for biofuel production (Makepa and Chihobo, 2024).

Finally, the environmental trade-offs must be carefully considered. Removing too much residue from fields reduces soil organic carbon, undermining climate mitigation and soil sustainability efforts. In addition, the high moisture content of many residues reduces their energy density and complicates handling, storage, and transport. Drying methods, including solar drying, can improve calorific value and ease of processing, but these require added investment and infrastructure. Balancing these trade-offs is crucial if residue-based biofuels are to serve as a sustainable and scalable component of the global energy system.

Despite the challenges, crop residues present immense opportunities for advancing sustainable bioenergy. One of the foremost advantages is the contribution to energy security. Fossil fuel reserves are finite and subject to geopolitical instability, which makes countries that rely heavily on imports vulnerable to supply disruptions and price fluctuations. Utilizing agricultural residues as feedstocks for biofuels can reduce this dependency, providing a locally available and renewable energy source. This is particularly important for energy-deficient nations that struggle with the economic burden of expensive fuel imports, as biofuels derived from residues could substantially ease their energy bills while strengthening national resilience (Hiloidhari et al., 2014).

Another major opportunity lies in GHG mitigation. Although the conversion of biomass into fuels requires some fossil energy input, the overall lifecycle emissions of biofuels are significantly lower than those of petroleum-derived fuels. Substituting fossil fuels with bioethanol, biodiesel, or biogas from crop residues can reduce net carbon emissions, mitigate air pollution, and contribute to climate change goals. Studies have shown that replacing even a portion of conventional diesel or gasoline with biofuels can cut CO_2 emissions by several kilograms per liter of fuel displaced, highlighting their potential as an effective tool for decarbonization (Khanna et al., 2011).

Biofuel production also provides avenues for rural development and job creation. Since the collection, processing, and conversion of residues are labor-intensive, they can generate additional income streams for farmers and employment opportunities for rural communities. The establishment of decentralized biogas plants, bioethanol distilleries, or biodiesel units can stimulate local economies, reduce rural-to-urban migration, and empower smallholder farmers by giving value to resources that are otherwise considered waste.

In addition to energy benefits, crop residues serve as a renewable feedstock resource for the chemical industry, offering a sustainable alternative to petrochemicals. Lignin, cellulose, and hemicellulose fractions can be converted into biochemicals, polymers, and advanced materials, positioning bioenergy within a broader biorefinery framework. This integrated approach can improve economic viability by diversifying outputs beyond fuels, creating additional value from the same biomass input.

Another promising opportunity is the favorable energy balance of certain biofuel pathways. For instance,. bioethanol production from crops such as switchgrass can yield several times more renewable energy than the fossil energy required for its production. Similarly, residues such as sugarcane bagasse or corn stover show strong energy return ratios, making them highly efficient bioenergy sources (Keshwani and Cheng, 2009). While biofuels may not fully replace fossil fuels, they represent a crucial part of the renewable energy mix, contributing to diversification and resilience of supply.

Lastly, second-generation biofuels offer a way to address the long-debated food-versus-fuel issue. By utilizing nonfood crops and agricultural residues, bioenergy development avoids direct competition with food production and can even complement agricultural systems. Some crops, such as sweet sorghum, provide both food (grain) and fuel (bagasse), enabling dual benefits. Furthermore, the cultivation of nonedible crops on marginal lands helps prevent pressure on prime agricultural areas while supporting sustainable land management practices. When combined with appropriate residue management strategies, biofuels can simultaneously enhance energy supply, safeguard soil health, and strengthen agricultural productivity (Kumar et al., 2024).

1.12 Conclusion

Agricultural residues represent a vast, renewable, and underexploited resource with significant potential to support global bioenergy goals. Their utilization through diverse conversion pathways not only reduces reliance on fossil fuels but also mitigates environmental impacts and enhances rural development. Nevertheless, large-scale deployment is constrained by economic, technical, and ecological challenges, particularly regarding residue collection, pretreatment costs, and soil health preservation. A balanced approach—one that integrates sustainable harvesting practices, efficient conversion technologies, and supportive policy frameworks—is essential to unlock their potential. By valorizing agricultural residues in integrated biorefineries, it is possible to achieve both energy generation and value-added coproducts, thereby fostering a resilient and circular bioeconomy.

References

Albuquerque, M.M., de Bona Sartor, G., Martinez-Burgos, W.J., Scapini, T., Edwiges, T., Soccol, C.R., Medeiros, A.B.P., 2024. Biohydrogen produced via dark fermentation: a review. Methane. 2674-0389. 3 (3), 500–532. doi:10.3390/methane3030029.

Aleminew, A., 2024. Enhancing soil fertility and crop productivity through crop residue management: a review. Agricultural Reviews, 0976-0741. (Of). doi:10.18805/ag.rf-294.

Ali, M., Saleem, M., Khan, Z., Watson, I.A., 2019. The Use of Crop Residues for Biofuel Production Biomass, Biopolymer-Based Materials, and Bioenergy: Construction, Biomedical, and Other Industrial Applications. Elsevier, Pakistan, pp. 369–395. Available from http://www.sciencedirect.com/science/book/9780081024263. doi:10.1016/B978-0-08-102426-3.00016-3.

Ariyawansha, T., Abeyrathna, D., Ahamed, T., Noguchi, R., 2022. Integrated bagasse utilization system based on hydrothermal liquefaction in sugarcane mills: theoretical approach compared with present practices. Biomass Convers. Biorefin. 2190-6815. 12 (1), 27–37. doi:10.1007/s13399-020-00958-w.

Asaad, S.M., Tawalbeh, M., Ali, A., Kindi, S.R.A., Al-Othman, A., 2023. Definition of Bioenergy Renewable Energy - Volume 2: Wave, Geothermal, and Bioenergy Definitions, Developments, Applications, Case Studies, and Modelling and Simulation.

Elsevier, United Arab Emirates, pp. 215–243. Available from https://www.sciencedirect.com/book/9780323952118. doi:10.1016/B978-0-323-95211-8.00004-X.

Azadbakht, M., Ardebili, S.M.S., Rahmani, M., 2023. A study on biodiesel production using agricultural wastes and animal fats. Biomass Convers. Biorefin. 21906823. 13 (6), 4893–4899. Available from https://www.springer.com/journal/13399. doi:10.1007/s13399-021-01393-1.

Blanco-Canqui, H., Lal, R., 2009. Crop residue removal impacts on soil productivity and environmental quality. Crit. Rev. Plant Sci. 15497836. 28 (3), 139–163. doi:10.1080/07352680902776507.

Cadenas, A., Cabezudo, S., 1998. Biofuels as sustainable technologies: perspectives for less developed countries. Technol. Forecast. Soc. Change. 00401625. 58 (1-2), 83–103. Available from https://www.journals.elsevier.com/technological-forecasting-and-social-change. doi:10.1016/s0040-1625(97)00083-8.

Caicedo, M., Barros, J., Ordás, B., 2016. Redefining agricultural residues as bioenergy feedstocks. Materials. 1996-1944. 9 (8), 635. doi:10.3390/ma9080635.

Chowdhury, P., Mahi, N.A., Yeassin, R., Chowdhury, N.-U.-R., Farrok, O., 2025. Biomass to biofuel: impacts and mitigation of environmental, health, and socioeconomic challenges. Energy Convers. Manag. X. 25901745. 25, 100889. doi:10.1016/j.ecmx.2025.100889.

Dizaji, H.B., Zeng, T., Hartmann, I., Enke, D., Schliermann, T., Lenz, V., Bidabadi, M., 2019. Generation of high quality biogenic silica by combustion of rice husk and rice straw combined with pre- and post-treatment strategies—a review. Appl. Sci. 2076-3417. 9 (6), 1083. doi:10.3390/app9061083.

Egwuatu, D.C., Ogunye, R.O., Anene, P.C., Alhassan, Y., Nwafor, I.R., Udoh, J.I., Asenuga, O.O., Nwokafor, C.V., 2024. A review on conversion of agricultural waste to bioenergy: processes and environmental impacts. Biotechnol. J. Int. 2456-7051. 28 (6), 18–34. doi:10.9734/bji/2024/v28i6746.

Hassan, M.K., Chowdhury, R., Ghosh, S., Manna, D., Pappinen, A., Kuittinen, S., 2021. Energy and environmental impact assessment of Indian rice straw for the production of second-generation bioethanol. Sustain. Energy Technol. Assess. 22131388. 47, 101546. doi:10.1016/j.seta.2021.101546.

Hiloidhari, M., Das, D., Baruah, D.C., 2014. Bioenergy potential from crop residue biomass in India. Renew. Sustain. Energy Rev. 13640321. 32, 504–512. Available from https://www.journals.elsevier.com/renewable-and-sustainable-energy-reviews. doi:10.1016/j.rser.2014.01.025.

Hitam, C.N.C., Jalil, A.A., 2023. A review on biohydrogen production through photo-fermentation of lignocellulosic biomass. Biomass Convers. Biorefin. 2190-6815. 13 (10), 8465–8483. doi:10.1007/s13399-020-01140-y.

Honorato-Salazar, J.A., Sadhukhan, J., 2020. Annual biomass variation of agriculture crops and forestry residues, and seasonality of crop residues for energy production in Mexico. Food Bioprod. Process. 09603085. 119, 1–19. Available from http://www.elsevier.com/wps/find/journaldescription.cws_home/713883/description#description. doi:10.1016/j.fbp.2019.10.005.

Jenkins, B.M., Sumner, H.R., 1986. Harvesting and handling agricultural residues for energy. Trans. Am. Soc. Agric. Eng. 29 (3), 824–836.

Kemausuor, F., Kamp, A., Thomsen, S.T., Bensah, E.C., Østergård, H., 2014. Assessment of biomass residue availability and bioenergy yields in Ghana. Resources, Conservation and Recycling. 09213449. 86, 28–37. doi:10.1016/j.resconrec.2014.01.007.

Keshwani, D.R., Cheng, J.J., 2009. Switchgrass for bioethanol and other value-added applications: a review. Bioresour. Technol. 09608524. 100 (4), 1515–1523. doi:10.1016/j.biortech.2008.09.035.

Khanna, M., Crago, C.L., Black, M., 2011. Can biofuels be a solution to climate change? The implications of land use change-related emissions for policy. Interface Focus. 20428901.

1 (2), 233–247. Available from http://rsfs.royalsocietypublishing.org/content/1/2/233.full.pdf+html. doi:10.1098/rsfs.2010.0016.

Kumar, B., Singh, M., Kumar, A., Rana, R., Kumar, S., Kumar, R., Singh, A.K., 2024. Sweet sorghum utilization for dual-purpose: feedfodder and fuel production (a review). Ecology, Environment and Conservation. 0971765X. 30 (02), 852–857. doi:10.53550/eec.2024.v30i02.072.

Kumar, R., Singh, S., Singh, O.V., 2008. Bioconversion of lignocellulosic biomass: biochemical and molecular perspectives. J. Ind. Microbiol. Biotechnol. 14765535. 35 (5), 377–391. doi:10.1007/s10295-008-0327-8.

Lackner, M., Besharati, M., 2025. Agricultural waste: challenges and solutions, a review. Waste. 2813-0391. 3 (2), 18. doi:10.3390/waste3020018.

Liu, X., Hoekman, S.K., Broch, A., 2017. Potential water requirements of increased ethanol fuel in the USA. Energy, Sustainability and Society. 21920567. 7 (1). Available from http://www.energsustainsoc.com/. doi:10.1186/s13705-017-0121-4.

Loy, A.C.M., Gan, D.K.W., Yusup, S., Chin, B.L.F., Lam, M.K., Shahbaz, M., Unrean, P., Acda, M.N., Rianawati, E., 2018. Thermogravimetric kinetic modelling of in-situ catalytic pyrolytic conversion of rice husk to bioenergy using rice hull ash catalyst. Bioresour. Technol. 18732976. 261, 213–222. Available from https://www.elsevier.com/locate/biortech. doi:10.1016/j.biortech.2018.04.020.

Makepa, D.C., Chihobo, C.H., 2024. Barriers to commercial deployment of biorefineries: a multi-faceted review of obstacles across the innovation chain. Heliyon. 24058440. 10 (12). Available from https://www.sciencedirect.com/science/journal/24058440. doi:10.1016/j.heliyon.2024.e32649.

Medina-Morales, M.A., De la Cruz-andrade, L.E., Paredes-Peña, L.A., Morales-Martínez, T.K., Rodríguez-de la Garza, J.A., Moreno-Dávila, I.M., Tamayo-Ordóñez, M.C., Rios-González, L.J., 2021. Biohydrogen production from thermochemically pretreated corncob using a mixed culture bioaugmented with Clostridium acetobutylicum. Int. J. Hydrogen Energy. 03603199. 46 (51), 25974–25984. Available from http://www.journals.elsevier.com/international-journal-of-hydrogen-energy/. doi:10.1016/j.ijhydene.2021.04.046.

Mlonka-Mędrala, A., Evangelopoulos, P., Sieradzka, M., Zajemska, M., Magdziarz, A., 2021. Pyrolysis of agricultural waste biomass towards production of gas fuel and high-quality char: experimental and numerical investigations. Fuel. 00162361. 296, 120611. doi:10.1016/j.fuel.2021.120611.

Muh, E., Tabet, F., Amara, S., 2021. Biomass conversion to fuels and value-added chemicals: a comprehensive review of the thermochemical processes. Curr. Altern. Energy. 24054631. 4 (1), 3–25. doi:10.2174/2405463103666191022121648.

Ntaikou, I., Antonopoulou, G., Lyberatos, G., 2010. Biohydrogen production from biomass and wastes via dark fermentation: a review. Waste Biomass Valorization. 1877-2641. 1 (1), 21–39. doi:10.1007/s12649-009-9001-2.

Pandey, B.K., Mishra, S., Dhar, R., Srivastava, R., 2023. Biological Hydrogen Production Driven by Photo-fermentation Processes Solar-Driven Green Hydrogen Generation and Storage. Elsevier, India, pp. 223–235. Available from https://www.sciencedirect.com/book/9780323995801. doi:10.1016/B978-0-323-99580-1.00015-7.

Pathak, V.V., Kothari, R., 2022. Crop Residues as a Potential Substrate for Bioenergy Production: An Overview. Informa UK Limited, pp. 121–138. doi:10.1201/9781003363231-6.

Policastro, G., Giugliano, M., Luongo, V., Napolitano, R., Fabbricino, M., 2022. Enhancing photo fermentative hydrogen production using ethanol rich dark fermentation effluents. Int. J. Hydrogen Energy. 03603199. 47 (1), 117–126. doi:10.1016/j.ijhydene.2021.10.028.

Reid, W.V., Ali, M.K., Field, C.B., 2020. The future of bioenergy. Glob. Change Biol. 13652486. 26 (1), 274–286. Available from http://onlinelibrary.wiley.com/journal/10.1111/(ISSN)1365-2486. doi:10.1111/gcb.14883.

Same, N.N., Yakub, A.O., Chaulagain, D., Tangoh, A.F., Nsafon, B.E.K., Owolabi, A.B., Suh, D., Huh, J.-S., 2024. The future of clean energy: agricultural residues as a bioethanol source and its ecological impacts in Africa. Renew. Energy. 09601481. 237, 121612. doi:10.1016/j.renene.2024.121612.

Sánchez-Arreola, E., Martin-Torres, G., Lozada-Ramírez, J.D., Hernández, L.R., Bandala-González, E.R., Bach, H., 2015. Biodiesel production and de-oiled seed cake nutritional values of a Mexican edible Jatropha curcas. Renew. Energy. 09601481. 76, 143–147. doi:10.1016/j.renene.2014.11.017.

Sarangi, P.K., Nayak, M.M., 2021. Agro-Waste for Second-Generation Biofuels Liquid Biofuels: Fundamentals, Characterization, and Applications. wiley, India, pp. 697–709. Available from https://onlinelibrary.wiley.com/doi/book/10.1002/9781119793038. doi:10.1002/9781119793038.ch20.

Saravanan, A., Ragini, Y.P., Karishma, S., Hemavathy, R.V., Jyotsna, M., 2025. A review on advancing sustainable energy: the role of biomass and bioenergy in a circular economy. Sustain. Futures. 26661888. 10, 100835. doi:10.1016/j.sftr.2025.100835.

Shahzad, H.M.A., Asim, Z., Khan, S.J., Almomani, F., Mahmoud, K.A., Mustafa, M.R.U., Rasool, K., 2024. Thermochemical and biochemical conversion of agricultural waste for bioenergy production: an updated review. Discover Environ. 27319431. 2 (1). Available from https://link.springer.com/journal/44274. doi:10.1007/s44274-024-00171-w.

Slade, R., Bauen, A., Gross, R., 2014. Global bioenergy resources. Nat. Clim. Change. 17586798. 4 (2), 99–105. doi:10.1038/nclimate2097.

Strzalka, R., Schneider, D., Eicker, U., 2017. Current status of bioenergy technologies in Germany. Renew. Sustain. Energy Rev. 18790690. 72, 801–820. Available from https://www.journals.elsevier.com/renewable-and-sustainable-energy-reviews. doi:10.1016/j.rser.2017.01.091.

Torquati, B., Marino, D., Venanzi, S., Porceddu, P., Chiorri, M., 2016. Using tree crop pruning residues for energy purposes: a spatial analysis and an evaluation of the economic and environmental sustainability. Biomass Bioenergy. 09619534. 95, 124–131. doi:10.1016/j.biombioe.2016.09.017.

Vaez, S., Karimi, K., Denayer, J.F.M., Kumar, R., 2023. Evaluation of apple pomace biochemical transformation to biofuels and pectin through a sustainable biorefinery. Biomass Bioenergy. 09619534. 172, 106757. doi:10.1016/j.biombioe.2023.106757.

Wang, L., Wang, H., Fan, J., Han, Z., 2023. Synthesis, catalysts and enhancement technologies of biodiesel from oil feedstock – a review. Sci. Total Environ. 00489697. 904, 166982. doi:10.1016/j.scitotenv.2023.166982.

Yanchal Jugwanth, Y.S.-S., Kana, E.B.G., 2020. Valorization of sugarcane bagasse for bioethanol production through simultaneous saccharification and fermentation: optimization and kinetic studies. Fuel. 00162361. 262, 116552. doi:10.1016/j.fuel.2019.116552.

You, S., Yong Sik, O., Tsang, D.C.W., Kwon, E.E., Wang, C.-H., 2018. Towards practical application of gasification: a critical review from syngas and biochar perspectives. Crit. Rev. Environ. Sci. Technol. 1064-3389. 48 (22-24), 1165–1213. doi:10.1080/10643389.2018.1518860.

Zambare, V., Patankar, R., Bhusare, B., Christopher, L., 2021. Recent advances in feedstock and lipase research and development towards commercialization of enzymatic biodiesel. Processes. 2227-9717. 9 (10), 1743. doi:10.3390/pr9101743.

Zhao, L., Wang, Z., Ren, H.-Y., Chen, C., Nan, J., Cao, G.-L., Yang, -S.-S., Ren, N.-Q., 2021. Residue cornstalk derived biochar promotes direct bio-hydrogen production from anaerobic fermentation of cornstalk. Bioresour. Technol. 09608524. 320 (Pt A), 124338. doi:10.1016/j.biortech.2020.124338.

CHAPTER 2

Agricultural residues as a feedstock for bioenergy production

Deepa Thomas[a], Reshmy Rajasekharan[b], Rekha Unni[c], Aravind Madhavan[d], Rahul Krishna Bhuvanendran[e], and Raveendran Sindhu[e,f]

[a]Post Graduate and Research Department of Chemistry, Bishop Moore College, Mavelikara, Kerala, India
[b]Cellupro Green Pvt. Ltd., Industrial Development Area, Mavelikara, Kerala, India
[c]Post Graduate Department of Chemistry, Christian College, Chengannur, Kerala, India
[d]School of Biotechnology, Amrita Vishwa Vidyapeetham, Amritapuri, Kollam, Kerala, India
[e]Department of Food Technology, TKM Institute of Technology, Kollam, Kerala, India
[f]Santhigiri Scientific and Industrial Research Institute (SSIRI), Trivandrum, Kerala, India

2.1 Introduction

The global demand for sustainable and renewable energy sources has grown significantly in recent years, driven by the pressing need to mitigate climate change, reduce dependency on fossil fuels, and ensure energy security. Among the various renewable energy options, bioenergy has emerged as a promising solution due to its potential to harness energy from biological materials, often referred to as biomass. One of the most promising and underutilized sources of biomass is agricultural residues. These residues possess significant potential as feedstock for bioenergy production, offering a sustainable pathway to convert waste into valuable energy products such as biofuels, biogas, and bioelectricity.

Utilizing agricultural residues for bioenergy production presents multiple environmental, economic, and social benefits. Environmentally, it offers a means of waste reduction, decreases greenhouse gas (GHG) emissions, and promotes a circular economy by reusing agricultural by-products. Economically, it creates new revenue streams for farmers, reduces costs associated with waste management, and stimulates rural development. From a social perspective, the bioenergy sector can generate employment opportunities and foster energy access in remote and underserved areas (Chen, 2014).

Despite its advantages, the use of agricultural residues for bioenergy production also faces several challenges. Issues such as collection, storage,

Bioenergy from Agricultural Residues
https://doi.org/10.1016/B978-0-443-24726-2.00004-3

transportation, and conversion efficiency must be addressed to ensure the feasibility and sustainability of this approach. Advances in technology, policy support, and investment in infrastructure are crucial to overcoming these barriers. This chapter explores the role of agricultural residues as a feedstock for bioenergy production, highlighting the key types of residues, conversion technologies, and the environmental, economic, and social impacts of their utilization. By leveraging these untapped resources, society can move closer to a more sustainable and low-carbon energy future.

2.2 Types and sources of agricultural residues

Agricultural residues are the major sources for bioenergy production and can be categorized as primary residues, secondary residues, and tertiary residues.

2.2.1 Primary residues/field residues

Primary residues in bioenergy production refer to the biomass materials that are generated as byproducts during the harvesting and processing of primary products such as crops, forestry, or other natural resources. These residues are considered a sustainable source of energy because they utilize materials that would otherwise go unused or be discarded. They are generated directly from agricultural fields, forests, or similar primary production activities. These residues are generally classified based on their origin: Agricultural residues, forestry residues, and aquatic biomass residues (Cai et al., 2017; Stefanidis et al., 2014).

These are the leftover parts of plants after harvesting the primary product like grains, fruits, or vegetables. These are in the form of straw (e.g., wheat, rice, barley); corn stover (stalks, leaves, and husks); sugarcane bagasse (crushed cane fibre after juice extraction); pruning and cuttings (branches and leaves from orchard crops and vineyards); and root and tuber residues (stems and roots from crops such as cassava or potatoes). Forestry residues include, namely, logging residues (materials left after timber harvesting, such as small branches), tree tops, bark, and forest floor debris. The harvested wastes from algae or aquatic plants are also used for bioenergy production.

2.2.2 Secondary residues

Secondary residues are byproducts left after the industrial or commercial processing of primary resources such as agricultural products, forestry products, or organic materials. They are typically rich in organic content and can be repurposed to produce bioenergy through various technologies. Secondary residues are categorized based on their origin or industrial sector.

Agricultural and food processing residues include oilseed processing residues such as oilseed cakes or meals left after oil extraction from seeds such as soybean, rapeseed, or sunflower. Fruit and vegetable processing waste such as pulp, peels, and seeds generated during juice extraction or canning processes are other major secondary residues. Grain milling byproducts such as bran, husks, or germ left after flour or starch production and forestry and wood processing residues, namely sawdust, wood chips, and black liquor containing lignin, are also active sources of bioenergy. Another class of secondary residues includes livestock and dairy industry residues often used in biodiesel production. Manure residues include processed animal waste rich in organic matter for anaerobic digestion. Urban and industrial organic waste for bioenergy is mainly food waste from restaurants or markets.

2.2.3 Tertiary residues

Tertiary residues refer to waste materials derived from postconsumer or postindustrial activities, and they can serve as valuable feedstocks for bioenergy production. These residues typically include materials that would otherwise be disposed of, such as municipal solid waste, construction and demolition debris, and discarded packaging (Yadav et al., 2024b). Utilizing tertiary residues for bioenergy not only reduces waste but also contributes to sustainable energy production. Some examples of these residues are municipal solid waste such as nonrecyclable plastics, food waste, paper, and textiles. These materials can be converted into energy via incineration, anaerobic digestion, or gasification. Waste from construction and demolition, used cooking oils and grease, compostable plastics and paper packaging, and textile waste are also utilized as effective bioenergy sources.

2.3 Lignocellulosic composition of agricultural wastes

Agricultural wastes, comprising crop residues and animal manure, represent an abundant and renewable resource for the generation of sustainable bioenergy and bioproducts. Understanding the complexities of their lignocellulosic composition is essential to maximizing their potential. Agricultural waste, primarily composed of lignocellulosic biomass, boasts a complex molecular framework consisting of cellulose, hemicellulose, and lignin. Cellulose is a linear polymer of glucose molecules, while hemicellulose is a branched polymer of various sugars. Lignin, on the other hand, is a complex, three-dimensional polymer that provides structural support to the plant cell wall. Additionally, minor constituents such as ash and protein are present in trace amounts, exerting negligible influence on the biomass's composition, pretreatment, and conversion processes (Chen, 2014). The suitability of these wastes for bioconversion into different products is determined by their complex matrix of cellulose, hemicellulose, and lignin. Crop type, growing conditions, and harvesting methods are some of the variables that affect the lignocellulosic content of agricultural wastes, which varies greatly. For example, the cellulose content of corn stover and sugarcane bagasse is larger than that of wheat straw and rice straw. In contrast to other agricultural wastes, cow manure has a reduced lignocellulosic concentration (Cai et al., 2017). Table 2.1 describes the lignocellulosic composition of agricultural wastes. To enable efficient conversion, the intricate biomass matrices must be broken down into simpler building blocks. Interestingly, agricultural waste typically contains a higher proportion of cellulose compared to hemicellulose and lignin (Stefanidis et al., 2014).

Cellulose-rich biomass can be effectively fermented to produce ethanol, while biomass with a high lignin concentration can be pyrolyzed to produce bio-oil. Therefore, understanding the lignocellulosic composition of agricultural wastes is essential for developing targeted bioconversion strategies.

To sum up, agricultural waste is an important resource for the development of sustainable bioenergy and bioproducts. The development of novel technologies that effectively transform these wastes into useful goods will require additional study in order to lessen waste disposal problems and advance a more circular bioeconomy.

2.4 Pretreatment methods for agricultural residues

The accumulation of agricultural residues throughout the production cycle presents a substantial opportunity for harnessing sustainable biomass energy. However, to unlock their full potential, these agro-waste biomasses must undergo transformative cycles. A pivotal step in this sequence is the pretreatment of biomass, which facilitates the degradation of lignin and consequently amplifies the efficiency and productivity of bioenergy generation (Kumar et al., 2023). The most widely used pretreatment techniques are

Table 2.1 Lignocellulosic composition of agricultural wastes.

Agricultural Waste	Cellulose (%)	Hemicellulose (%)	Lignin (%)	References
Sugarcane bagasse	40-45	30-35	20-25	Yadav et al. (2024b)
Rice straw	28-32	23-27	12-15	Kadam et al. (2000)
Corn stover	38-40	28-30	17-20	Zhao et al. (2018)
Wheat straw	30-35	25-30	15-20	Tian et al. (2018)
Cotton gin trash	30-35	25-30	20-25	Haque et al. (2021)
Coconut husk	35-40	25-30	20-25	Din et al. (2021)
Sunflower stalks	35-40	25-30	15-20	Ziebell et al. (2013)
Barley straw	32-35	22-25	12-15	Borrega et al. (2022)
Soybean hulls	35-40	25-30	15-20	Bittencourt et al. (2021)
Peanut hulls	30-35	20-25	10-15	Kerr et al. (1986)
Banana leaves	20-25	15-20	10-15	Fernandes et al. (2013)
Pineapple leaves	25-30	20-25	15-20	Saini et al. (2022)
Sorghum straw	30-35	25-30	15-20	Wahyuni et al. (2019)
Hemp stalks	30-35	25-30	15-20	Chaowana et al. (2024)
Tobacco stalks	25-30	20-25	10-15	Hu et al. (2024)
Canola straw	10-15	20-25	10-15	Huerta and Saldaña (2019)
Flax straw	28-32	22-25	12-15	Hailemariam and Woldeyes (2024)

chemical, biological, physicochemical, physical, and green solvent-based methods (Fig. 2.1).

2.4.1 Physical pretreatment methods

2.4.1.1 Mechanical pretreatment

The mechanical size reduction of lignocellulosic biomass through methods like chipping, grinding, shearing, or milling reduces particle size and increases surface area, allowing cellulolytic enzymes to penetrate the biomass more easily and boosting cellulose conversion rates (Mosier et al., 2005).

Mechanical pretreatment offers several advantages, including the absence of inhibitor production, simplicity of operation, and significant reduction in substrate size, which enhances surface area and improves

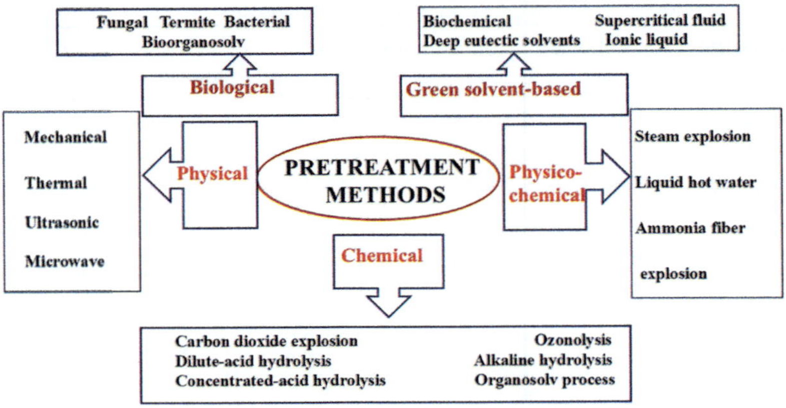

Fig. 2.1 Pretreatment methods.

biomethane production from wheat straw. In addition, it reduces crystallinity, increases porosity, and enhances enzyme digestibility. However, this method also has several drawbacks, including high operational costs, high energy consumption, and an inability to degrade lignin, which limits its effectiveness. Furthermore, mechanical pretreatment often requires combination with other pretreatment methods and results in low sugar yields, impacting overall efficiency (Zhao et al., 2022).

2.4.1.2 Thermal pretreatment

The application of thermal energy to lignocellulosic biomass facilitates the disintegration of complex polysaccharides, such as xylan, araban, and mannan. This thermal pretreatment enhances the susceptibility of cellulose and hemicellulose to enzymatic hydrolysis, thereby augmenting the production of valuable biochemicals. Although this method offers numerous benefits, including environmental sustainability and reliability, it also poses challenges related to the formation of inhibitory compounds and the requirement for elevated temperatures. Nonetheless, thermal pretreatment has demonstrated considerable potential in optimizing the conversion of recalcitrant biomass into valuable products (Solikhin et al., 2017).

2.4.1.3 Ultrasonic pretreatment

Ultrasonic pretreatment is an innovative method that utilizes low-energy ultrasound waves to disrupt the intricate structure of lignocellulosic biomass. This method offers several advantages, including reduced energy consumption, minimal inhibitor formation, and low waste generation. Moreover, ultrasonic pretreatment has been found to enhance hydrogen fermentation from bagasse and facilitate the production of multiple valuable products. However, this technique also presents some challenges, such as low selectivity, energy loss in diluted media, and difficulties in scaling up reactor design. Nonetheless, ultrasonic pretreatment holds promise for intensifying biomass processing and improving the overall efficiency of biofuel production (Vu et al., 2020).

2.4.1.4 Microwave pretreatment

Microwave pretreatment employs electromagnetic radiation to permeate the matrix of agricultural waste, disrupting the intermolecular bonds that hold its constituent parts together. This nonthermal process selectively targets the lignocellulosic structure, facilitating the breakdown of complex biomass components. The advantages of microwave pretreatment include rapid processing times, uniform heating patterns, energy efficiency, and minimal generation of inhibitory compounds. However, this method also has some drawbacks, such as high capital investment costs, limited scalability, potential for uneven heating patterns, and the need for precise control of microwave frequency and intensity (Nande et al., 2023).

2.4.2 Physicochemical pretreatment methods

2.4.2.1 Steam explosion

Steam explosion is a pretreatment method that involves exposing lignocellulosic biomass to saturated steam at temperatures ranging from 160°C to 290°C and pressures of 0.69–4.85 MPa for several seconds or minutes. The steam is then rapidly decompressed to atmospheric pressure, causing the biomass to explode and resulting in a significant reduction in size with relatively low energy input. This method can achieve 80–100% hydrolysis of hemicellulose, destruction of a portion of the xylan fraction, and 45–65% recovery of xylose. However, steam explosion can also lead to the formation of inhibitors, which can be mitigated by adding chemicals such as $H2SO_4$, SO_2, or CO_2 to improve the efficiency of subsequent enzymatic hydrolysis and cellulose depolymerization (Mosier et al., 2005).

2.4.2.2 Liquid hot water

LHW pretreatment utilizes pressurized hot water at temperatures between 170°C and 230°C and pressures up to 5 MPa for 1–46 min, with solids loads below 20%. By applying temperatures between 170°C and 230°C and pressures up to 5 MPa for extended periods, LHW effectively disrupts the biomass structure, solubilizing up to 50% of the lignin and achieving nearly complete hemicellulose hydrolysis. Additionally, this method partially solubilizes lignin (20–50%), making it valuable for lignin valorization. LHW pretreatment also offers low energy requirements, minimal chemical usage, and scalability. Moreover, LHW can be readily integrated with other pretreatment techniques to further enhance its effectiveness. Overall, LHW pretreatment represents a highly promising strategy for efficient biomass conversion and lignin valorization, offering a compelling solution for the production of sustainable biofuels and biochemicals (Ballesteros et al., 2002).

2.4.2.3 Ammonia fiber explosion

LHW pretreatment is a thermochemical process that utilizes pressurized hot water to break down lignocellulosic biomass. Operating at temperatures between 170°C and 230°C, pressures below 5 MPa, and treatment times ranging from 1 to 46 min, LHW can effectively process biomass with solids loads below 20%. This method yields several key benefits, including the redistribution of lignin, hydrolysis of 80–100% of hemicellulose, and recovery of 88–98% of xylose. Moreover, LHW minimizes the formation of inhibitors, converts up to 90% of cellulose into fermentable sugars, and partially solubilizes lignin (20–50%) for potential valorization (Koegel et al., 1999).

The advantages of LHW pretreatment extend beyond its technical benefits. This method boasts low energy requirements compared to other pretreatment methods, minimizing environmental impact and operating costs through reduced chemical usage. In addition, LHW is scalable and flexible, accommodating various biomass feedstocks and process conditions. Its compatibility with downstream biological and chemical processes, as well as its potential for integration with other pretreatment methods, further underscores its effectiveness (Ibrahim et al., 2018).

2.4.3 Chemical pretreatment methods

2.4.3.1 Carbon dioxide explosion

The carbon dioxide explosion technique is a pioneering approach in biomass pretreatment, leveraging the potent disruptive capabilities of high-pressure CO_2 to deconstruct the intricate architecture of lignocellulosic biomass. This innovative method involves injecting a precisely calibrated amount of

CO_2–4 kg per kg of fiber, at a pressure of 5.62 MPa, thereby inducing a rapid and efficient breakdown of the biomass matrix. Notably, CO2 explosion enables up to 75% cellulose conversion, making it an attractive approach for enhancing enzymatic hydrolysis and fermentation (Sun and Cheng, 2002). This method offers several additional benefits, including low energy requirements, minimal chemical usage, and reduced environmental impact. Moreover, this versatile technique can be readily integrated with existing biomass processing technologies, yielding a highly scalable, economically viable, and environmentally sustainable solution for the production of advanced biofuels and specialty biochemicals (Jain and Kumar, 2024).

2.4.3.2 Ozonolysis

Ozonolysis is a chemical pretreatment method that utilizes ozone to break down lignocellulosic biomass at room temperature and pressure. By leveraging ozone's unique reactivity, this approach effectively dismantles the recalcitrant lignin matrix, achieving impressive degradation yields of up to 57%. Consequently, ozonolysis unlocks the cellulose component, facilitating enhanced conversion rates of up to 57% without generating inhibitory compounds that can impede downstream biological processes. This method's mild operating conditions, minimal energy requirements, and lack of chemical additives make it an attractive and environmentally friendly option for biomass pretreatment. Moreover, this versatile methodology can be seamlessly integrated with other pretreatment techniques to create synergistic effects, further optimizing biomass conversion and valorization (Jönsson and Martín, 2016).

2.4.3.3 Dilute-acid hydrolysis

Dilute-acid hydrolysis is a widely employed pretreatment technique that leverages the catalytic properties of diluted acids, such as H2SO4, HCl, or HNO3, to break down lignocellulosic biomass. This versatile process can be operated in continuous or batch modes, accommodating a broad range of solids loads and temperatures. This versatile process can be operated in continuous or batch modes, accommodating a broad range of solids loads and temperatures. While continuous processing is better suited for lower solids loads, batch processing is more effective for higher loads. A key consideration in this method is the need for pH neutralization, resulting in gypsum generation. Nonetheless, dilute-acid hydrolysis achieves remarkable hydrolysis yields, including near-complete hemicellulose breakdown and substantial xylose recovery. Furthermore, elevated temperatures during this process favor further cellulose hydrolysis, while lignin is redistributed without being solubilized, thereby enhancing its accessibility for subsequent processing steps (Niju et al., 2019).

2.4.3.4 Concentrated-acid hydrolysis

Concentrated-acid hydrolysis is a pretreatment method that employs high concentrations of sulfuric acid (10–30%) at elevated temperatures (170–190°C) to break down lignocellulosic biomass. This process is characterized by a distinct solid-liquid ratio of 1:1.6 and can be augmented by the incorporation of peracetic acid (21–60%) within a silo-type reactor configuration. However, this method necessitates acid recovery due to the high acid concentrations used. In addition, the residence time required for concentrated-acid hydrolysis is longer compared to dilute-acid hydrolysis. Furthermore, the required residence time for concentrated-acid hydrolysis exceeds that of dilute-acid hydrolysis, introducing additional process design and operational complexities. Notably, the use of peracetic acid induces lignin oxidation, which can impact subsequent processing steps (Rezania et al., 2020).

2.4.3.5 Alkaline hydrolysis

Alkaline hydrolysis is a gentle yet potent pretreatment strategy that harnesses the selective deconstruction capabilities of dilute alkaline solutions, comprising sodium hydroxide (NaOH) or calcium hydroxide ($Ca(OH)_2$), to efficiently break down lignocellulosic biomass. This versatile approach can be finely tuned to meet specific process requirements through deliberate variations in temperature, reaction time, and the judicious addition of hydrogen peroxide (H_2O_2) as a precision oxidizing agent. Notably, alkaline hydrolysis offers several advantages, including lower reactor costs compared to acid-based pretreatment methods. The hydrolysis efficiency of this approach is impressive, with 50% hemicellulose breakdown, 60–75% xylose recovery, and minimal inhibitor formation. Furthermore, alkaline hydrolysis induces cellulose swelling, enhancing its susceptibility to enzymatic hydrolysis, and can achieve further cellulose conversion yields of up to 65%. In addition, this method facilitates the removal of a significant proportion of lignin, ranging from 24 to 55% for hardwood and lower percentages for softwood, thereby enhancing the overall accessibility of biomass components for subsequent conversion processes and paving the way for the production of high-value bio-based products (Xu et al., 2020).

2.4.3.6 Organosolv process

The Organosolv process represents a paradigm shift in biomass pretreatment, harnessing the potent solvent capabilities of organic compounds such as methanol, ethanol, and acetone, often augmented by a minor acidic component (1% H_2SO_4 or HCl). This process operates at moderate temperatures (185–198°C) and residence times (30–60 min) and a fixed pH range (2.0–3.4). While solvent recovery is an essential aspect of this process, the organosolv method yields an impressive array of benefits, including exhaustive hemicellulose hydrolysis, exceptional xylose yields, and a nearly complete dissolution and fragmentation of lignin, as well as the cleavage of internal lignin-hemicellulose bonds (Rabelo et al., 2023).

2.4.4 Biological

2.4.4.1 Fungal pretreatment

Fungal pretreatment represents a novel, microbially-mediated biomass deconstruction to leverage the enzymatic capabilities of various fungal species, including brown-, white-, and soft-rot fungi. Through solid-state fermentation, these microorganisms produce a suite of enzymes, comprising cellulases, hemicellulases, and lignin-degrading enzymes, such as ligninases, lignin peroxidases, polyphenol oxidases, laccases, and quinone-reducing enzymes. This enzymatic consortium collaborates to dismantle the intricate lignocellulosic matrix, liberating fermentable sugars and other valuable biomass constituents. Despite its relatively protracted timeline, fungal pretreatment offers a compelling alternative to conventional chemical-based methods, distinguished by its benign environmental footprint, moderate operating conditions, and diminished energy requirements, rendering it an attractive strategy for sustainable biomass processing (Kainthola et al., 2021).

2.4.4.2 Bioorganosolv pretreatment

Bioorganosolv pretreatment represents a synergistic, eco-friendly approach that combines the selective lignin-degrading capabilities of the fungus *Ceriporiopsis subvermispora* with the solvent properties of ethanol. A 2-h ethanolysis phase at 140–200°C comes after the initial 2–8 week fungal treatment in this two-stage procedure. The fungal decomposition of the lignin network facilitates subsequent hemicellulose hydrolysis, which is achieved through the action of ethanol. The potential for ethanol reuse, a 15% decrease in the amount of electricity needed for ethanolysis, and an environmentally friendly procedure are just a few benefits of this biological pretreatment approach (Rabelo et al., 2023).

2.4.4.3 Termite pretreatment

Termite pretreatment stands out for its distinctive benefits, which encompass mild operating conditions, reduced energy consumption, and a remarkable lignin degradation efficiency exceeding 60%. In addition, this approach is characterized by its environmental sustainability and economic feasibility. Nevertheless, termite pretreatment also has its drawbacks, notably a prolonged duration and relatively limited efficiency (Yadav et al., 2024a).

2.4.4.4 Bacterial pretreatment

Bacterial pretreatment presents a promising, eco-friendly approach to lignocellulosic biomass conversion, characterized by low energy consumption and the bioconversion of lignocellulose into biomethane and other biofuels. This method also produces minimal hazardous chemicals, making it an environmentally friendly option. However, bacterial pretreatment requires a longer incubation time, which can be a drawback (Ferdeş et al., 2020).

2.4.5 Green solvent-based pretreatment

2.4.5.1 Biochemical pretreatment

Green solvent-based biochemical pretreatment offers a highly efficient method for biomass conversion, as evidenced by its successful application in converting wheat straw to biomethane. This strategy has a number of benefits, such as an exceptionally brief pretreatment period and no inhibitor production, which could impede subsequent procedures. This strategy has a number of benefits, such as an exceptionally brief pretreatment period and no inhibitor production, which could impede subsequent procedures (Baksi et al., 2023).

2.4.5.2 Ionic liquid pretreatment

Ionic liquid pretreatment is a highly effective method for converting lignocellulosic biomass to bioethanol, boasting a remarkable 99% reliability. A broad temperature range, strong chemical and thermal stability, and a rapid recovery rate are only a few benefits of this method. Nevertheless, there are a number of serious disadvantages to ionic liquid treatments. These include excessive energy consumption, toxicity, and unaffordable cost. Furthermore, ionic liquids exhibit strong intermolecular interactions, are difficult to formulate, and possess low volatility. In addition, they cannot be reused, exhibit low biodegradability, and are highly viscous, ultimately limiting their practical applications (Usmani et al., 2020).

2.4.5.3 DES pretreatment

DES pretreatment emerges as a promising, economical approach for improving the biodegradability of lignocellulosic biomass, including sugarcane bagasse, poplar wood, and bamboo. This method boasts several advantages, including high biodegradability, low toxicity, and high recovery rates. A sustainable substitute for conventional energy sources, DES pretreatment also makes it easier to produce biofuels from lignocellulosic biomass. Even though DES has a high viscosity, its many advantages make it an attractive option for biomass pretreatment and subsequent biofuel production (Chen and Mu, Chen and Tiancheng, 2019).

2.4.5.4 Supercritical fluid pretreatment

Supercritical fluid pretreatment, utilizing carbon dioxide (CO_2), presents a promising approach for enhancing biomass bioconversion. This method offers several advantages, including low viscosity, good diffusivity, improved sugar yields from sorghum straw, and digestibility of green coconut fiber. In addition, supercritical CO_2 is environmentally friendly, nonflammable, and recyclable. Despite the fact that this process uses a lot of energy owing to the need for high temperatures, pressures, and specialized equipment, it has the advantage of employing relatively cheap CO_2. Furthermore, the use of supercritical CO_2 preserves the native biomass structure, making it an attractive option for sustainable biomass processing (Daza Serna et al., 2016).

2.5 Conversion technologies for bioenergy production

Waste materials are converted into a variety of fuels that can be used to generate energy as part of the waste-to-energy conversion process. Many different technologies have been developed and used to turn waste into bioenergy. These technologies are broadly classified into thermochemical and biochemical methods. Fig. 2.2 represents various types of biomass conversion technologies.

2.5.1 Thermochemical methods

Thermochemical conversion is the technique of utilizing heat to convert the breakdown of organic materials. Liquid fuels have been produced from biomass using a variety of thermochemical processes, such as direct combustion, torrefaction, hydrothermal liquefaction, pyrolysis, gasification, and incineration. In practice, the biomass breaks down under controlled operating circumstances to create solid, liquid, and gas (syngas), which require an

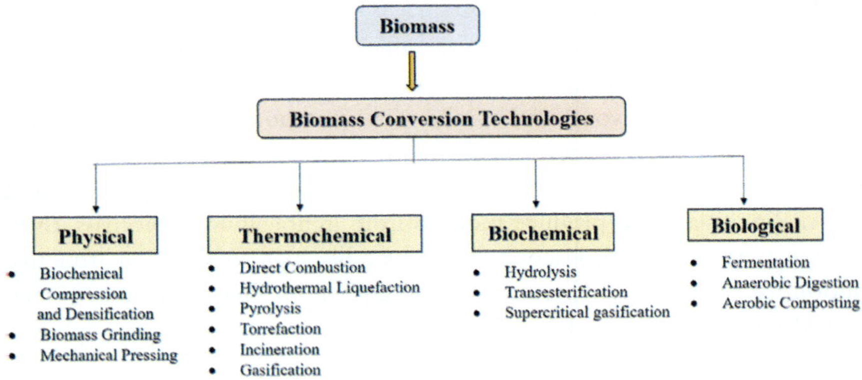

Fig. 2.2 Types of biomass conversion technologies.

additional catalytic stimulation procedure to make liquid fuels known as drop-in biofuel. One of its key advantages is that, in contrast to biochemical conversion techniques, it may use any kind of biomass as a feedstock for biomaterials Fig. 2.2.

2.5.1.1 *Direct combustion*

Direct combustion for biomass to bioenergy refers to the process of burning biomass in the presence of oxygen to produce heat, which can then be used for power generation, heating, or industrial processes. It is one of the simplest and most widely used methods for converting biomass into energy. The main processes involved are feedstock preparation, combustion (Kadam et al., 2000; Zhao et al., 2018).

2.5.1.2 *Torrefaction*

Torrefaction is an endothermic pretreatment process that mostly takes place in an oxygen-free environment at temperatures between 200 and 300°C with a nonaccelerated heating rate of <50°C/min. Through this process, solid biomass is upgraded to create a torrefied product that can subsequently be utilized as a viable coal substitute. The torrefaction process can be broadly divided into three modes of operation: dry, moist, and steam.

Biomaterial feedstock can undergo dry torrefaction in either an oxidative or nonoxidative environment. Air, flue gases, and different oxygen concentrations between 200 and 300°C are utilized as carrier gases in oxidative media, while nitrogen and carbon dioxide are utilized in nonoxidative atmospheres. The biomass can also be torrefied in a moist atmosphere, usually with water and diluted acids, at 180–260°C for 5–240 min. The material that is generated is called hydrochar. In addition to the two previously described methods, biomass is torrefied by steam torrefaction, which involves the introduction of steam at high temperatures and pressures. The biomass is torrefied at 200–260°C and 5–10 min for environmental temperature and contact time, respectively. Steam will be able to bulge and divide the biomass with minimal loss in the feedstock due to the subsequent fast venting of the pressure (Tian et al., 2018; Haque et al., 2021).

2.5.1.3 Hydrothermal liquefaction

A thermochemical process known as hydrothermal liquefaction is used to effectively break down lignocellulosic feedstock, whether it is dry or wet, into renewable liquid fuel. It can be separated into two primary subclasses based on how it operates: solvent liquefaction and indirect liquefaction. The first case involves the conversion of biomass or its liquefied constituents into syngas, which is then followed by the production of fuel (alkanes and alcohols). On the other hand, in the second, proton solvents such as alcohols, water, phenols, and ionic liquids directly convert biomass into liquid fuel. Due to its exceptional benefits, favorable operating circumstances, and increased product yield, solvent liquefaction should be used first rather than the alternative.

In this process, the wet biomass is typically used directly, reducing the need for energy-intensive drying processes required in other thermochemical methods. The crude biomass is mixed with water to form slurry. The biomass-water mixture is subjected to high temperatures (200–350°C) and high pressures (10–25 MPa). At these conditions, water acts as a reactant, solvent, and catalyst, facilitating the breakdown of organic molecules. Biomass polymers such as cellulose, hemicellulose, lignin, and proteins decompose into smaller molecules. During the process, procedures such as hydrolysis, depolymerization, and repolymerization occur, forming bio-crude oil, gases, and other by-products. The reaction mixture is cooled, and the products are separated into different phases, namely bio-crude oil, aqueous phase, gaseous products (CO_2, H_2, and CH_4), and solid residue (unreacted feedstock and minerals as ash).

Furthermore, owing to its superior physicochemical characteristics, hydrothermal liquefaction is thought to be a more efficient method with lower energy usage than pyrolysis. Usually higher than those obtained from the traditional pyrolysis technique, the bio-crude produced by the hydrothermal liquefaction process has an oxygen content of 10–20 wt percent and a thermal value of 30–35 MJ/kg (Din et al., 2021; Ziebell et al., 2013).

2.5.1.4 Pyrolysis

Pyrolysis is a widely used thermochemical process for breaking down carbonaceous biomass, including cellulose, hemicellulose, and lignin. Through an endothermic reaction, solid, liquid, and gas biofuels are produced in an oxygen-free environment. The method involves drying of biomass followed by chopping, grounding, or pulverizing to ensure uniform heating and reaction rates. Then it is heated to the desired pyrolysis temperature (300–800°C) in the absence of oxygen. The decomposition temperatures of different components vary with temperature, such as hemicelluloses (200–300°C), cellulose (300–400°C), and lignin (250–500°C). Volatile compounds are released, leaving behind a solid residue (biochar).

The thermal breakdown of biomass involves a number of complex procedures, including dehydration, decarboxylation, decarbonylation, hydrogenation, isomerization, aromatization, depolymerization, and charring. There are several types of pyrolysis, including slow, rapid, catalytic, microwave-assisted, carbon dioxide, additive, solar, and hydro-pyrolysis.

Slow pyrolysis is a thermochemical process in which biomass is heated slowly in an oxygen-deficient environment to produce biochar as the primary product, along with smaller amounts of bio-oil and syngas. This method focuses on maximizing the solid residue (biochar), making it ideal for applications requiring carbon-rich materials for energy, soil improvement, or carbon sequestration. The igniting temperature ranges from 300 to 500°C, the heating rate is low (0.1–2°C/s), and the biomass residence time is long (hours to days). A technique known as "fast pyrolysis" involves rapidly heating organic materials at speeds greater than 103°C s^{-1} and contacting them for shorter periods of time—up to three seconds—without the presence of air. The main product is oil in liquid form. Fast pyrolysis, also known as ultrafast pyrolysis, is a rapidly accelerated pyrolysis process with significant gas and oil products at a high heating rate. The shortest dwell time, fastest heating rate, and medium-to-high temperature (700–1000°C) are the working parameters. Liquid condensate (10–20%), gases (60–80%), and char (10–15%) are the output yields (Borrega et al., 2022; Amaro Bittencourt et al., 2021).

Catalytic pyrolysis is an advanced thermochemical process that integrates a catalyst into the pyrolysis process to improve the quality and selectivity of the products, particularly bio-oil. The use of catalysts helps break down biomass more effectively, reducing oxygen content in bio-oil, minimizing by-products, and enhancing the production of hydrocarbons and valuable chemicals. The types of catalysts in pyrolysis are zeolites—highly porous aluminosilicates used for cracking and deoxygenation; metal oxides (e.g., Al_2O_3, TiO_2, MgO); alkali metal salts (e.g., K_2CO_3, Na_2CO_3); and bifunctional catalysts—combine multiple functionalities (e.g., acidity for cracking and metal sites for hydrogenation).

Microwave-assisted pyrolysis is an advanced thermochemical process in which biomass is heated using microwave radiation rather than conventional heating methods. The unique heating mechanism of microwaves allows for rapid, uniform, and efficient thermal decomposition of biomass, leading to the production of bio-oil, syngas, and biochar. Microwaves generate heat by directly interacting with the biomass or a microwave-absorbing material (susceptor) mixed with the biomass. These materials convert microwave energy into thermal energy, enabling precise temperature control and faster heating. Carbon dioxide (CO_2)-assisted pyrolysis is a thermochemical process that utilizes CO_2 as a reactant or process medium to convert biomass into valuable products such as biochar, bio-oil, and syngas. The integration of CO_2 introduces unique chemical interactions that enhance the efficiency of the pyrolysis process, improve the quality of products, and enable effective carbon utilization. The processes involved in CO_2-assisted pyrolysis are biomass (e.g., agricultural residues, wood, or algae) preprocessed (drying and size reduction) for uniform thermal conversion. Then a steady supply of CO_2 is ensured, often sourced from industrial emissions or storage. The reactor is heated to operating temperatures, typically 400–900°C. CO_2 is introduced into the reactor either continuously or batch-wise. Biomass is thermally broken down into volatiles, biochar, and gases. CO_2 interacts with the carbon in biomass to promote secondary reactions such as gasification. The products obtained after CO_2-assisted pyrolysis are vapors that can be condensed into bio-oil and noncondensable gases (enriched with CO and H_2) that are collected as syngas. Biochar is separated for collection and further applications (Kerr et al., 1986).

Additive-assisted pyrolysis is a thermochemical process in which additives are introduced to modify or enhance the pyrolysis reaction, product yield, and quality. Additives can include catalysts, chemicals, or inert materials that influence the thermal decomposition of biomass or feedstock, directing the process toward specific outcomes. The various types of additives used in pyrolysis are zeolites, metal oxides (CaO and MgO), inorganic salts (KCl or NaCl), acidic or basic additives, carbonaceous additives (activated carbon), and nanomaterials (Ni and Fe).

Solar pyrolysis is an innovative approach to thermochemical biomass conversion that uses concentrated solar energy as the primary heat source.

This sustainable process minimizes reliance on fossil fuels by directly utilizing sunlight to drive the pyrolysis reactions, producing biochar, bio-oil, and syngas. The key components of solar pyrolysis involve solar concentrators, reactor design, and thermal energy storage methods.

2.5.1.5 Gasification

Gasification is a thermochemical process that converts organic materials (like biomass) into a combustible gas called "syngas" (synthetic gas) under high temperatures (typically 700–1000°C) in a controlled oxygen-limited environment. Unlike combustion, which uses excess oxygen to burn biomass completely, gasification uses limited oxygen, leading to partial oxidation. This syngas can be used as a fuel to produce electricity, heat, or chemicals, making it a versatile method for bioenergy production (Fernandes et al., 2013). The process of gasification involves four primary stages: drying (Temperature: ~100-200°C), pyrolysis (Temperature: ~200-500°C), reduction (Temperature: ~800-1000°C), and oxidation (Temperature: ~700-1000°C). These steps produce CO_2 and H_2O and release heat for endothermic reactions.

The main products of gasification are syngas (synthetic gas contains CO, H_2, CH_4, CO_2, H_2O, and small amounts of other gases). The syngas can be used as a fuel for power generation (electricity production via gas engines or gas turbines), a feedstock for chemical production (e.g., methanol, ammonia, and synthetic natural gas), or a fuel for combustion (for heating or direct burning). Biochar is a solid carbon-rich residue that can be used as a soil amendment (for carbon sequestration and soil fertility improvement) or as a fuel for heating. Tar and condensable hydrocarbons are by-products that can create operational issues (such as clogging) but can be collected and used as liquid fuels or chemicals. Ash is inorganic material that remains as solid residue, which can be used as a construction material or disposed of safely.

The advantages of gasification are low emissions of pollutants (such as NOx, SOx, and particulates) and high energy recovery potential due to syngas utilization. The limitations are tar formation, which can clog the system and reduce efficiency. Gasification plants are more complex and expensive than combustion systems. Advanced technology and skilled operators are required for optimal performance (Saini et al., 2022).

2.5.1.6 Incineration

Incineration is the process of oxidative burning at high temperatures (900–1100°C) in an aerobic environment with the goal of recovering energy and reducing volume. Ash, solids, and combustion gas are the end products of the incineration process. The three most crucial stages in the incineration process are gasification, pyrolysis, evaporation, and degassing. However, the use of incineration pollutes the air and water since it produces a lot of chemicals, hazardous gasses, and GHG emissions. Therefore, the establishment of an incinerator facility in any location will require both local community support and innovative, efficient emission-cleaning techniques (Wahyuni et al., 2019).

2.5.2 Biological methods

Bioenergy production involves converting biological materials (biomass) into energy, such as electricity, heat, or biofuels. The key biological methods rely on the metabolic activities of microorganisms, plants, and enzymes to break down biomass and convert it into usable energy. Processes of biological conversion are thought to be environmentally benign. However, they are time-consuming due to their fermentation length, the usage of expensive hydrolytic enzymes, and the sensitivity of microbes to many factors such as growth conditions, contaminants, sugar substrates, nutrients, and inhibitors. Anaerobic digestion, aerobic composting, and fermentation are a few biological techniques.

2.5.2.1 Anaerobic digestion

Anaerobic digestion is a biological process in which microorganisms break down organic material in an oxygen-free environment, producing biogas [mainly methane (CH_4) and carbon dioxide (CO_2)] and digestate (a nutrient-rich byproduct). This method is widely used to produce renewable energy from organic waste, making it a key technology in waste management and sustainable energy production (Chaowana et al., 2024). The anaerobic digestion process occurs in four main stages, each driven by a specific group of microorganisms: hydrolysis, acidogenesis, acetogenesis, and methanogenesis.

2.5.2.2 Aerobic composting

Aerobic composting is a biological process where microorganisms decompose organic materials in the presence of oxygen. While composting is primarily known as a method for producing organic fertilizers, it also plays a role in bioenergy production by generating heat energy. Unlike anaerobic digestion, composting does not produce methane (CH_4) or other combustible gases. However, the heat produced during the process can be harnessed for energy applications. Aerobic composting occurs in three main stages and is driven by a diverse microbial community (Hu et al., 2024; Huerta and Saldaña, 2019). The mesophilic phase (20–40°C), thermophilic phase (40–70°C), and maturation phase (cooling and curing) are the three phases of aerobic composting. The temperature gradually decreases as the easily degradable materials are exhausted. The compost stabilizes, and humus-like material is formed.

2.5.2.3 Fermentation

The fermentation method for aerobic composting involves the breakdown of organic material by microorganisms in the presence of oxygen. While "fermentation" is often associated with anaerobic (oxygen-free) processes such as alcohol production, in composting, aerobic microorganisms (bacteria, fungi, and actinomycetes) play a vital role. The term "fermentation" in this context refers to the microbial metabolic processes that decompose complex organic material into simpler substances while producing heat, CO_2, water vapor, and stabilized organic matter (compost) (Hailemariam and Woldeyes, 2024).

During aerobic composting, microorganisms break down organic material through oxidative metabolism rather than anaerobic fermentation. The process is facilitated by the continuous supply of oxygen and proper management of moisture, temperature, and pH. The following stages are involved, namely initial breakdown (hydrolysis), oxidation of organic compounds, thermophilic phase (heat fermentation), and finally stabilization and humification.

2.5.3 Biochemical methods

Hydrolysis, transesterification, and supercritical water gasification are examples of biochemical conversion methods that combine chemical and biological processes. This process turns biomass into gas (CO_2/CH_4), waste (manure or compost), and water (H_2O) using microbes and biological catalysts. It offers excellent selectivity in converting biomass into the desired end products (Kumar et al., 2023).

2.5.3.1 Hydrolysis

Hydrolysis is a key process in bioenergy production, particularly in the conversion of biomass into simpler, fermentable sugars. These sugars are subsequently converted into biofuels such as bioethanol, biobutanol, and biogas through processes such as fermentation or anaerobic digestion. Hydrolysis plays a crucial role in the breakdown of complex carbohydrates, proteins, and fats into smaller, simpler molecules that can be used as energy feedstocks. Hydrolysis is a chemical reaction where water molecules (H_2O) break down larger, complex biomolecules into simpler components. Enzymes (biocatalysts) speed up this reaction. The main goal of hydrolysis in bioenergy production is to convert biomass (which contains carbohydrates, proteins, and lipids) into simpler forms, such as glucose, amino acids, and fatty acids (Mosier et al., 2005).

Types of hydrolysis in bioenergy production include carbohydrate hydrolysis (breakdown of cellulose and starch), protein hydrolysis (breakdown of proteins into amino acids), and lipid hydrolysis (breakdown of fats into fatty acids and glycerol).

2.5.3.2 Transesterification

Transesterification is a chemical process used to convert oils and fats (triglycerides) into biodiesel and glycerol. It is one of the most important processes in biodiesel production, which is a key part of the bioenergy sector. The process can be catalyzed using acids, bases, or enzymes (biocatalysts), with alkaline-catalyzed transesterification being the most common industrial method. Base catalysts widely used are sodium hydroxide (NaOH) and potassium hydroxide (KOH); acid catalysts such as sulfuric acid (H_2SO_4) or hydrochloric acid (HCl); and enzymatic catalysts (Biocatalysts) such as lipase enzymes (Zhao et al., 2022).

2.5.3.3 Supercritical water gasification

Supercritical water gasification is an advanced method of converting wet biomass into a gaseous fuel rich in hydrogen (H_2), methane (CH_4), carbon monoxide (CO), and carbon dioxide (CO_2). This process occurs under supercritical water conditions—when water is heated above its critical temperature (374°C) and critical pressure (22.1 MPa)—giving it unique properties that enhance the conversion of biomass into valuable gases (Solikhin et al., 2017).

2.5.4 Physical methods

Physical methods of bioenergy production typically involve mechanical processes that convert biomass into usable forms of energy without the need for chemical reactions or biological processes. These methods primarily focus on converting biomass into solid fuels, liquid fuels, or gaseous fuels through compression, grinding, densification, or other physical treatments. Physical methods can play an essential role in bioenergy

Table 2.2 Advantages and disadvantages of bioenergy production technologies.

Technology	Advantages	Disadvantages	References
Thermochemical	Utilizes diverse biomass to rapidly produce syngas, bio-oil, biochar, heat, and electricity.	High initial investment and operating costs, pollutant emissions	Mosier et al. (2005)
Biological	Environmentally friendly, simple and low-cost technology,	Slow process, large land and space requirements	Ballesteros et al. (2002)
Biochemical	Renewable and sustainable, low greenhouse gas emissions	Slow process, feedstock limitations, high water demand	Kumar et al. (2023)
Physical	Simple and low-cost process, reduces biomass volume, short processing time	Limited energy conversion, limited product versatility	Vu et al. (2020)

generation by improving the energy density of biomass, making it easier to handle, transport, and store (Vu et al., 2020). The main physical methods of bioenergy production are biomass compression and densification through pelletizing, briquetting, or cubing and biomass grinding and mechanical pressing (Physical oil extraction) (Nande et al., 2023). The advantages and disadvantages of various bioenergy technologies can be listed in Table 2.2.

2.6 Conclusion and future perspectives

It is an obvious fact that agricultural residues can give maximum biomass to make possible bioenergy production. It can bring about a circular economy, as well as reduce dependence on fossil fuels. Actually, waste from agriculture can treat many issues environmentally, economically, and socially in making valuable bioenergy products. Progresses in technologies for the conversion of biomass, improved logistics, collection, and transport, as well as policy frameworks that support the initiatives must be present to make agricultural residues completely efficient and viable as sources of bioenergy in the future. Further investment in R&D and collaboration with various stakeholders would yield innovation and economies of scale for production and, hence, make a huge impact on global energy security and reduction efforts in climate change at the end.

References

Baksi, S., Saha, D., Saha, S., Sarkar, U., Basu, D., Kuniyal, J.C., 2023. Pre-treatment of lignocellulosic biomass: review of various physico-chemical and biological methods influencing the extent of biomass depolymerization. Int. J. Environ. Sci. Technol. 1735-1472. 20 (12), 13895–13922. doi:10.1007/s13762-023-04838-4.

Ballesteros, I., Oliva, J.M., Negro, M.J., Manzanares, P., Ballesteros, M., 2002. Ethanol production from olive oil extraction residue pretreated with hot water. Appl. Biochem. Biotechnol. 02732289. 98-100, 717–732. doi:10.1385/ABAB:98-100:1-9:717.

Bittencourt, G.A., de Souza Vandenberghe, L.P., Valladares-Diestra, K., Herrmann, L.W., de Mello, A.F.M., Vásquez, Z.S., Karp, S.G., Soccol, C.R., 2021. Soybean hulls as carbohydrate feedstock for medium to high-value biomolecule production in biorefineries: a review. Bioresour. Technol. 09608524. 339, 125594. doi:10.1016/j.biortech.2021.125594.

Borrega, M., Hinkka, V., Hörhammer, H., Kataja, K., Kenttä, E., Ketoja, J.A., Palmgren, R., Salo, M., Sundqvist-Andberg, H., Tanaka, A., 2022. Utilizing and valorizing oat and barley straw as an alternative source of lignocellulosic fibers. Materials. 19961944. 15 (21). Available from http://www.mdpi.com/journal/materials. doi:10.3390/ma15217826.

Cai, J., He, Y., Yu, X., Banks, S.W., Yang, Y., Zhang, X., Yu, Y., Liu, R., Bridgwater, A.V., 2017. Review of physicochemical properties and analytical characterization of lignocellulosic biomass. Renew. Sustain. Energy Rev. 18790690. 76, 309–322. Available from https://www.journals.elsevier.com/renewable-and-sustainable-energy-reviews. doi:10.1016/j.rser.2017.03.072.

Chaowana, P., Hnoocham, W., Chaiprapat, S., Yimlamai, P., Chitbanyong, K., Wanitpinyo, K., Chaisan, T., Paopun, Y., Pisutpiched, S., Khantayanuwong, S., Puangsin, B., 2024. Utilization of hemp stalk as a potential resource for bioenergy. Mater. Sci. Energy Technol. 25892991. 7, 19–28. Available from https://www.keaipublishing.com/en/journals/materials-science-for-energy-technologies/. doi:10.1016/j.mset.2023.07.001.

Chen, H., 2014. Chemical Composition and Structure of Natural Lignocellulose. In: Biotechnology of Lignocellulose, 25–71. doi:10.1007/978-94-007-6898-7_2.

Chen, Y., Tiancheng, M., 2019. Application of deep eutectic solvents in biomass pretreatment and conversion. Green Energy Environ. 24680257. 4 (2), 95–115. doi:10.1016/j.gee.2019.01.012.

Daza Serna, L.V., Alzate, C.E.O., Alzate, C.A.C., 2016. Supercritical fluids as a green technology for the pretreatment of lignocellulosic biomass. Bioresour. Technol. 09608524. 199, 113–120. doi:10.1016/j.biortech.2015.09.078.

Din, N.A.S., Lim, S.J., Maskat, M.Y., Zaini, N.A.M., 2021. Bioconversion of coconut husk fibre through biorefinery process of alkaline pretreatment and enzymatic hydrolysis. Biomass Convers. Biorefin. 21906823. 11 (3), 815–826. Available from http://www.springer.com/engineering/energy+technology/journal/13399. doi:10.1007/s13399-020-00895-8.

Ferdeş, M., Dincă, M.N., Moiceanu, G., Zabava, B.Ş., Paraschiv, G., 2020. Microorganisms and enzymes used in the biological pretreatment of the substrate to enhance biogas production: a review. Sustainability. 20711050. 12 (17). Available from https://res.mdpi.com/d_attachment/sustainability/sustainability-12-07205/article_deploy/sustainability-12-07205-v2.pdf. doi:10.3390/su12177205.

Fernandes, E.R.K., Marangoni, C., Souza, O., Sellin, N., 2013. Thermochemical characterization of banana leaves as a potential energy source. Energy Convers. Manag. 01968904. 75, 603–608. doi:10.1016/j.enconman.2013.08.008.

Hailemariam, T.T., Woldeyes, B., 2024. Production and characterization of pulp and paper from flax straw. Sci. Rep. 20452322. 14 (1). Available from https://www.nature.com/srep/. doi:10.1038/s41598-024-74096-y.

Haque, A.N.M.A., Remadevi, R., Naebe, M., 2021. A review on cotton gin trash: sustainable commodity for material fabrication. J. Clean. Prod. 09596526. 281. Available from https://www.journals.elsevier.com/journal-of-cleaner-production. doi:10.1016/j.jclepro.2020.125300.

Hu, N., Liu, X., Wei, S., Yao, J., Wang, W., Liu, B., Tang, T., Jiang, J., Wang, L., 2024. Current status and future prospects of pretreatment for tobacco stalk lignocellulose. Front. Bioeng. Biotechnol. 22964185. 12. Available from http://journal.frontiersin.org/journal/bioengineering-and-biotechnology#archive. doi:10.3389/fbioe.2024.1465419.

Huerta, R.R., Saldaña, M.D.A., 2019. Sequential treatment with pressurized fluid processing and ultrasonication for biorefinery of canola straw towards lignocellulosic nanofiber production. Ind. Crops Prod. 09266690. 139. Available from https://www.elsevier.com/inca/publications/store/5/2/2/8/2/5. doi:10.1016/j.indcrop.2019.111521.

Ibrahim, M.F., Kim, S.W., Abd-Aziz, S., 2018. Advanced bioprocessing strategies for biobutanol production from biomass. Renew. Sustain. Energy Rev. 18790690. 91, 1192–1204. Available from https://www.journals.elsevier.com/renewable-and-sustainable-energy-reviews. doi:10.1016/j.rser.2018.04.060.

Jain, S., Kumar, S., 2024. Advances and challenges in pretreatment technologies for bioethanol production: a comprehensive review. Sustain. Chem. Clim. Action. 27728269. 5, 100053. doi:10.1016/j.scca.2024.100053.

Jönsson, L.J., Martín, C., 2016. Pretreatment of lignocellulose: formation of inhibitory by-products and strategies for minimizing their effects. Bioresour. Technol. 18732976. 199, 103–112. Available from https://www.elsevier.com/locate/biortech. doi:10.1016/j.biortech.2015.10.009.

Kadam, K.L., Forrest, L.H., Jacobson, W.A., 2000. Rice straw as a lignocellulosic resource: collection, processing, transportation, and environmental aspects. Biomass Bioenergy 18 (5), 369–389. doi:10.1016/S0961-9534(00)00005-2.

Kainthola, J., Podder, A., Fechner, M., Goel, R., 2021. An overview of fungal pretreatment processes for anaerobic digestion: applications, bottlenecks and future needs. Bioresour. Technol. 09608524. 321, 124397. doi:10.1016/j.biortech.2020.124397.

Kerr, T.J., Windham, W.R., Woodward, J.H., Benner, R., 1986. Chemical composition and in-vitro digestibility of thermochemically treated peanut hulls. J. Sci. Food Agric. 10970010. 37 (7), 632–636. doi:10.1002/jsfa.2740370706.

Koegel, R.G., Sreenath, H.K., Straub, R.J., 1999. Alfalfa fiber as a feedstock for ethanol and organic acids. unchanged 105–115. doi:10.1007/978-1-4612-1604-9_10.

Kumar, J.A., Sathish, S., Prabu, D., Renita, A.A., Saravanan, A., Deivayanai, V.C., Anish, M., Jayaprabakar, J., Baigenzhenov, O., Hosseini-Bandegharaei, A., 2023. Agricultural waste biomass for sustainable bioenergy production: feedstock, characterization and pretreatment methodologies. Chemosphere. 18791298. 331. Available from https://www.elsevier.com/locate/chemosphere. doi:10.1016/j.chemosphere.2023.138680.

Mosier, N., Wyman, C., Dale, B., Elander, R., Lee, Y.Y., Holtzapple, M., Ladisch, M., 2005. Features of promising technologies for pretreatment of lignocellulosic biomass. Bioresour. Technol. 09608524. 96 (6), 673–686. Available from https://www.elsevier.com/locate/biortech. doi:10.1016/j.biortech.2004.06.025.

Nande, A., Chaudhari, V., Raut, S., Dhoble, S.J., 2023. Management of Agriculture Waste Materials: Challenges and Future Aspects 360-Degree Waste Management, Volume 1: Fundamentals, Agricultural and Domestic Waste, and Remediation. Elsevier, India, pp. 19–37. Available from https://www.sciencedirect.com/book/9780323907606. doi:10.1016/B978-0-323-90760-6.00011-4.

Niju, S., Swathika, M., Balajii, M., 2019. Pretreatment of Lignocellulosic Sugarcane Leaves and Tops for Bioethanol Production Lignocellulosic Biomass to Liquid Biofuels. Elsevier, India, pp. 301–324. Available from http://www.sciencedirect.com/science/book/9780128159361. doi:10.1016/B978-0-12-815936-1.00010-1.

Rabelo, S.C., Nakasu, P.Y.S., Scopel, E., Araújo, M.F., Cardoso, L.H., Costa, A.C.D., 2023. Organosolv pretreatment for biorefineries: current status, perspectives, and challenges. Bioresour. Technol. 18732976. 369. Available from https://www.elsevier.com/locate/biortech. doi:10.1016/j.biortech.2022.128331.

Rezania, S., Oryani, B., Cho, J., Talaiekhozani, A., Sabbagh, F., Hashemi, B., Rupani, P.F., Mohammadi, A.A., 2020. Different pretreatment technologies of lignocellulosic biomass for bioethanol production: an overview. Energy. 03605442. 199. Available from https://www.journals.elsevier.com/energy. doi:10.1016/j.energy.2020.117457.

Saini, R., Chen, C.W., Patel, A.K., Saini, J.K., Di Dong, C., Singhania, R.R., 2022. Valorization of pineapple leaves waste for the production of bioethanol. Bioengineering 9, 557–559. doi:10.3390/BIOENGINEERING9100557.

Solikhin, A., Hadi, Y.S., Massijaya, M.Y., Nikmatin, S., 2017. Morphological and chemo-thermal changes of oven-heat treated oil palm empty fruit bunch fibers during dry disk milling. unchanged. 09768432. 14 (1), 9–17. Available from http://www.springer.com/life+sciences/forestry/journal/13196. doi:10.1007/s13196-016-0182-6.

Stefanidis, S.D., Kalogiannis, K.G., Iliopoulou, E.F., Michailof, C.M., Pilavachi, P.A., Lappas, A.A., 2014. A study of lignocellulosic biomass pyrolysis via the pyrolysis of cellulose, hemicellulose and lignin. J. Anal. Appl. Pyrolysis. 01652370. 105, 143–150. doi:10.1016/j.jaap.2013.10.013.

Sun, Y., Cheng, J., 2002. Hydrolysis of lignocellulosic materials for ethanol production: a review. Bioresour. Technol. 09608524. 83 (1), 1–11. Available from https://www.elsevier.com/locate/biortech. doi:10.1016/S0960-8524(01)00212-7.

Tian, S.Q., Zhao, R.Y., Chen, Z.C., 2018. Review of the pretreatment and bioconversion of lignocellulosic biomass from wheat straw materials. Renew. Sustain. Energy Rev. 18790690. 91, 483–489. Available from https://www.journals.elsevier.com/renewable-and-sustainable-energy-reviews. doi:10.1016/j.rser.2018.03.113.

Usmani, Z., Sharma, M., Gupta, P., Karpichev, Y., Gathergood, N., Bhat, R., Gupta, V.K., 2020. Ionic liquid based pretreatment of lignocellulosic biomass for enhanced bioconversion. Bioresour. Technol. 09608524. 304, 123003. doi:10.1016/j.biortech.2020.123003.

Vu, H.P., Nguyen, L.N., Vu, M.T., Johir, M.A.H., McLaughlan, R., Nghiem, L.D., 2020. A comprehensive review on the framework to valorise lignocellulosic biomass as biorefinery feedstocks. Sci. Total Environ. 18791026. 743. Available from https://www.elsevier.com/locate/scitotenv. doi:10.1016/j.scitotenv.2020.140630.

Wahyuni, Y., Miyamoto, T., Hartati, H., Widjayantie, D., Windiastri, V.E., Sulistyowati, Y., Rachmat, A., Hartati, N.S., Ragamustari, S.K., Tobimatsu, Y., Nugroho, S., Umezawa, T., 2019. Variation in lignocellulose characteristics of 30 Indonesian sorghum (Sorghum bicolor) accessions. Ind. Crops Prod. 09266690. 142, 111840. doi:10.1016/j.indcrop.2019.111840.

Xu, L., Zhang, S.J., Zhong, C., Li, B.Z., Yuan, Y.J., 2020. Alkali-Based Pretreatment-Facilitated Lignin Valorization: a Review. Ind. Eng. Chem. Res. 59, 16923–16938. doi:10.1021/ACS.IECR.0C01456/ASSET/IMAGES/MEDIUM/IE0C01456_0006.GIF.

Yadav, P., Dixit, Y., Asnani, H., Sharma, A.K., 2024a. Exploration of ethanologenic bacteria from termite gut for bioethanol production. Biomass Convers. Biorefin. 1–14. doi:10.1007/S13399-024-05935-1/METRICS.

Yadav, P., Anu, S., Tiwari, K., Kumar, V., Singh, D., Kumar, S., Manisha, V.M., Singh, B., 2024b. Sugarcane bagasse: an important lignocellulosic substrate for production of enzymes and biofuels. Biomass Convers. Biorefin. 21906823. 14 (5), 6111–6142. Available from https://www.springer.com/journal/13399. doi:10.1007/s13399-022-02791-9.

Zhao, L., Sun, Z.F., Zhang, C.C., Nan, J., Ren, N.Q., Lee, D.J., Chen, C., 2022. Advances in pretreatment of lignocellulosic biomass for bioenergy production: challenges and perspectives. Bioresour. Technol. 18732976. 343. Available from https://www.elsevier.com/locate/biortech. doi:10.1016/j.biortech.2021.126123.

Zhao, Y., Damgaard, A., Christensen, T.H., 2018. Bioethanol from corn stover – a review and technical assessment of alternative biotechnologies. Prog. Energy Combust. Sci. 03601285. 67, 275–291. Available from https://www.journals.elsevier.com/progress-in-energy-and-combustion-science. doi:10.1016/j.pecs.2018.03.004.

Ziebell, A.L., Barb, J.G., Sandhu, S., Moyers, B.T., Sykes, R.W., Doeppke, C., Gracom, K.L., Carlile, M., Marek, L.F., Davis, M.F., Knapp, S.J., Burke, J.M., 2013. Sunflower as a biofuels crop: an analysis oflignocellulosic chemical properties. Biomass Bioenergy. 09619534. 59, 208–217. doi:10.1016/j.biombioe.2013.06.009.

CHAPTER 3

Biogas production from agricultural residues

A. Mohan[a], Kotteeswaran Santhanam[b], B. Bharathiraja[c], R. Praveen Kumar[d] and D. Gayathri[e],

[a]Department of Civil Engineering, Easwari Engineering College, Ramapuram, Chennai, Tamil Nadu, India
[b]Department of Civil Engineering, SRM Institute of Science and Technology, Ramapuram, Chennai, Tamil Nadu, India
[c]Department of Biotechnology, Easwari Engineering College, Chennai, Tamil Nadu, India
[d]Department of Biotechnology, Arunai Engineering College, Tiruvannamalai, Tamil Nadu, India
[e]Department of Computer Science and Engineering, Velammal Engineering College, Surapet, Chennai, India

3.1 Overview of anaerobic digestion process

Anaerobic digestion is a biological process where microorganisms break down organic matter in the absence of oxygen, producing biogas (a mixture mainly of methane and carbon dioxide) and digestate (a nutrient-rich residue). It occurs in four stages (Fig. 3.1).

3.1.1 Hydrolysis

Hydrolysis is the first and most crucial step in the anaerobic digestion process, during which complex organic matter such as carbohydrates, proteins, and fats is broken down into simpler, soluble compounds that microorganisms can readily utilize. In this stage, hydrolytic bacteria secrete extracellular enzymes: cellulases to degrade cellulose into simple sugars such as glucose, proteases to break proteins into amino acids, and lipases to split fats into glycerol and long-chain fatty acids (Abanades et al., 2022). These reactions transform insoluble agricultural residues, such as straw, husks, and crop stalks, into water-soluble molecules, enabling their subsequent fermentation in later stages of digestion. The efficiency of hydrolysis is particularly important for lignocellulosic materials, as slow or incomplete breakdown can significantly limit biogas production.

3.1.2 Acidogenesis

Acidogenesis is the second stage of anaerobic digestion, in which the simple, soluble compounds produced during hydrolysis—such as sugars, amino acids, and glycerol—are further broken down by acidogenic (fermentative)

Bioenergy from Agricultural Residues
https://doi.org/10.1016/B978-0-443-24726-2.00001-8

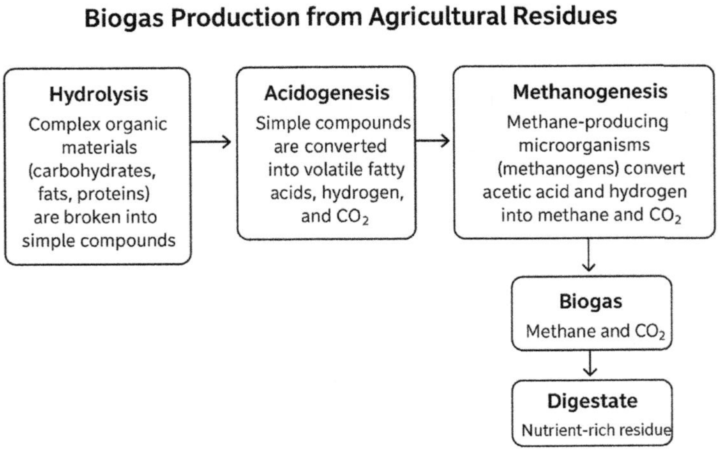

Fig. 3.1 Process of biogas production from agricultural residues.

bacteria. In this phase, the microorganisms convert these intermediates into volatile fatty acids (VFAs) such as acetic, propionic, and butyric acids, along with hydrogen (H_2), carbon dioxide (CO_2), alcohols, and other by-products (Abd et al., 2022). This stage is essentially a fermentation process, where the metabolic activity of the bacteria lowers the pH of the digester environment, making proper process control essential to avoid acid accumulation that can inhibit later stages. The VFAs and gases generated during acidogenesis serve as critical substrates for the next stage, acetogenesis, which further refines these products into compounds directly usable by methane-producing microorganisms.

3.1.3 Acetogenesis

Acetogenesis is the third stage of anaerobic digestion, during which the VFAs and alcohols generated in acidogenesis are converted into acetic acid, hydrogen (H_2), and carbon dioxide (CO_2) by acetogenic bacteria. These microorganisms operate under strictly anaerobic conditions and often in close association with hydrogen-consuming methanogens, as excessive hydrogen buildup can inhibit acetogenesis (Aboudi et al., 2019). Long-chain fatty acids, propionic acid, and butyric acid are oxidized into acetate: a key direct substrate for methane production in the final stage. This step acts as a critical biochemical bridge between the acidogenic phase, which produces a wide range of intermediate compounds, and methanogenesis, where these intermediates are transformed into energy-rich biogas. Maintaining favorable conditions for acetogenic bacteria, such as balanced pH and low hydrogen partial pressure, is essential for a stable and efficient digestion process.

3.1.4 Methanogenesis

Methanogenesis is the final stage of anaerobic digestion, in which methane-producing microorganisms, known as methanogens, convert acetic acid, hydrogen (H_2), and carbon dioxide (CO_2) into methane-rich biogas. This stage is carried out by strictly anaerobic archaea that thrive in oxygen-free environments and are highly sensitive to changes in temperature, pH, and toxic substances. Methanogenesis proceeds mainly through two biochemical pathways: acetoclastic methanogenesis, where acetate is split into methane and CO_2, and hydrogenotrophic methanogenesis, where hydrogen reduces CO_2 to methane. The efficiency of this stage largely determines the overall methane content of the biogas, typically ranging from 50% to 70%. Because methanogens grow relatively slowly compared to other microbial groups, maintaining stable digester conditions is essential to avoid process imbalances (Afridi et al., 2023). The methane generated here is the primary energy carrier of the biogas and can be used directly for heating, electricity generation, or upgraded to biomethane for use as a vehicle fuel.

3.2 Feedstock preparation and handling

Feedstock preparation and handling are critical steps in ensuring efficient biogas production from agricultural residues such as crop stalks, husks, straw, and animal manure. Since many of these materials, particularly lignocellulosic residues, have complex structures and low biodegradability, pretreatment is often necessary to enhance microbial access and digestion efficiency. Common preparation methods include size reduction through chopping or shredding to increase the surface area for enzymatic attack, and moisture adjustment to achieve the optimal solids concentration: typically 8%–12% for wet digestion systems (Ai et al., 2020). In addition, mixing and homogenization of feedstock help maintain uniform composition and prevent clogging or stratification within the digester. Removal of contaminants such as stones, plastics, and metals is essential to avoid mechanical damage and operational issues. For some residues, mild thermal, chemical, or biological pretreatments may be applied to break down lignin and hemicellulose, thereby accelerating hydrolysis in the digestion process. Proper storage and handling before feeding the digester also help preserve feedstock quality and minimize the loss of biodegradable material.

3.2.1 Size reduction

Size reduction, through methods such as chopping, shredding, or grinding, is an essential pretreatment step in preparing agricultural residues for anaerobic digestion. By breaking large particles into smaller fragments, the overall surface area available for microbial and enzymatic action is significantly increased, which accelerates the hydrolysis process and improves biogas yield. This step is particularly important for fibrous, lignocellulosic materials such as crop stalks, straw, and husks, where the dense structure can limit microbial penetration. Reduced particle size also promotes better mixing within the digester, prevents floating layers, and minimizes dead zones that can hinder digestion efficiency (Akbulut et al., 2021). However, the degree of size reduction must be balanced against the energy cost of mechanical processing to ensure overall process sustainability and economic viability.

3.2.2 Moisture adjustment

Moisture adjustment is a vital step in feedstock preparation, aimed at achieving the optimal solids content for efficient anaerobic digestion. In wet digestion systems, the total solids content is typically maintained between 8% and 12%, which provides favorable conditions for microbial activity, enhances mixing, and facilitates pumping within the digester. Agricultural residues such as straw, husks, and stalks often have low moisture content and require the addition of water, slurry, or liquid manure to reach the desired level. Conversely, materials with excessively high moisture may be blended with drier residues to balance the solids concentration. Proper moisture adjustment not only supports stable digestion but also helps prevent operational issues such as scum formation, sedimentation, or reduced biogas yield due to suboptimal microbial contact with the feedstock.

3.2.3 Mixing and homogenization

Mixing and homogenization are important pretreatment steps that ensure the feedstock entering the anaerobic digester has a uniform composition and consistency. Agricultural residues often vary in particle size, moisture content, and nutrient balance, which can lead to uneven digestion and reduced biogas yields if fed directly (Alengebawy et al., 2022c). By thoroughly blending different materials—such as crop residues with animal manure or liquid slurries—the feedstock achieves a more consistent texture, nutrient distribution, and moisture level. This uniformity promotes steady microbial activity, prevents the formation of floating or settled layers inside the digester, and minimizes the risk of localized over-acidification. In some cases, mechanical mixers or slurry pumps are used during storage to keep the material well mixed before it is fed into the digestion system. Effective homogenization ultimately leads to more stable operation and improved overall efficiency of the anaerobic digestion process.

3.2.4 Removal of contaminants

Removal of contaminants such as stones, plastics, and metals is a critical step in feedstock preparation for anaerobic digestion, as these foreign materials can damage equipment, hinder microbial activity, and reduce overall process efficiency. Agricultural residues collected from fields or storage areas often contain unwanted debris introduced during harvesting, handling, or transportation (Alengebawy et al., 2022a). Large, hard objects such as stones or metal fragments can cause mechanical wear or blockages in pumps and mixers, while plastics and other nonbiodegradable materials can accumulate in the digester, reducing active volume and necessitating costly cleaning operations. Screening, sieving, magnetic separation, or manual sorting are commonly employed methods to remove these impurities before feeding the material into the system. Ensuring contaminant-free feedstock not only protects the longevity of digestion equipment but also improves operational stability and the quality of both the biogas and the digestate produced.

3.3 Biogas yield and composition

Biogas yield is influenced by several factors, including the type and biodegradability of the feedstock, operating conditions within the digester, and the hydraulic retention time. Agricultural residues vary widely in their biogas potential: animal manure generally produces lower yields due to its partially digested nature, while crop residues rich in carbohydrates can generate higher volumes if properly pretreated. Optimal digestion conditions, such as maintaining mesophilic (35°C–40°C) or thermophilic (50°C–55°C) temperatures, stable pH, and adequate nutrient balance, are essential for maximizing methane production. Hydraulic retention time, typically ranging from 20 to 40 days in wet systems, ensures that microorganisms have sufficient contact with the substrate to achieve complete degradation (Alengebawy et al., 2022b). The composition of biogas is generally 50%–70% methane (CH_4), 30%–50% carbon dioxide (CO_2), and trace amounts of hydrogen sulfide (H_2S), ammonia (NH_3), and water vapor. A higher methane fraction increases the calorific value of the gas, making it more suitable for applications such as electricity generation, heating, or upgrading to biomethane for use as a vehicle fuel.

The composition of biogas produced from agricultural residues typically consists of methane (CH_4) in the range of 50%–70% by volume, which serves as the primary energy carrier due to its high calorific value. Carbon dioxide (CO_2) forms the second-largest component, usually accounting for 30%–50% of the total gas volume, and although it has no fuel value, it influences the combustion characteristics of the gas (Alengebawy et al., 2023). In addition to these main constituents, biogas contains trace amounts of other gases such as hydrogen sulfide (H_2S), ammonia (NH_3), and water vapor (Archana et al., 2024). Hydrogen sulfide, even in small concentrations, can be corrosive to equipment and must be removed during biogas upgrading, while ammonia can contribute to odor and potential inhibition of microbial activity if present in high levels. Water vapor is a natural by-product of the digestion process and can condense in pipelines, making proper gas drying or condensation management necessary. The exact composition depends on the feedstock characteristics and operating conditions, but maintaining a higher proportion of methane is desirable for improved energy performance.

The average biogas yield from agricultural residues varies significantly depending on the type of material and its organic content. Animal manure generally produces 20–40 m^3 of biogas per ton of fresh matter, as much of its easily degradable organic material has already been broken down in the animal's digestive system (Ardolino and Arena, 2019). In contrast, crop residues such as straw, husks, and stalks can yield between 100 and 300 m^3 of biogas per ton of volatile solids, provided they undergo appropriate pretreatment to enhance biodegradability. The actual yield depends on factors such as feedstock composition, particle size, digestion temperature, retention time, and process stability. Importantly, a higher methane content in the produced biogas directly increases its energy value, making process optimization for methane-rich gas an essential goal for efficient bioenergy production.

3.4 Digestate management and utilization

Digestate is the semi-solid residue that remains after the anaerobic digestion of agricultural residues. It contains the undigested organic matter along with a rich supply of nutrients such as nitrogen, phosphorus, and potassium, making it a valuable by-product for agricultural use (Arfan et al., 2023). Depending on the process design, digestate may be separated into solid and liquid fractions for easier handling and application. The solid portion can be composted or used directly as an organic soil conditioner, improving soil structure, water retention, and long-term fertility. The liquid fraction, rich in soluble nutrients, can be applied as a bioslurry or used in fertigation systems, reducing the need for synthetic fertilizers (Aryal et al., 2020). Proper digestate management is essential to avoid issues such as odor, pathogen spread, or nutrient runoff into water bodies. Storage facilities, application timing, and adherence to environmental regulations play a key role in ensuring its safe and sustainable use. In addition to agricultural benefits, digestate can also be further processed for specialized products such as biobased fertilizers, contributing to a circular bioeconomy.

3.4.1 Nutrient-rich composition of digestate

Digestate is inherently nutrient-rich, containing significant amounts of nitrogen, phosphorus, and potassium: the three primary macronutrients essential for plant growth. Nitrogen in digestate is often present in readily available forms such as ammonium, allowing for quick uptake by crops. Phosphorus contributes to root development and energy transfer within plants, while potassium supports water regulation, enzyme activation, and overall plant health (Atelge et al., 2020). In addition to these macronutrients, digestate also contains micronutrients such as calcium, magnesium, and sulfur, as well as organic matter that improves soil structure and microbial activity. This nutrient profile makes digestate a valuable substitute or supplement for chemical fertilizers, promoting sustainable agriculture while recycling the nutrients originally contained in the feedstock materials (Fig. 3.2).

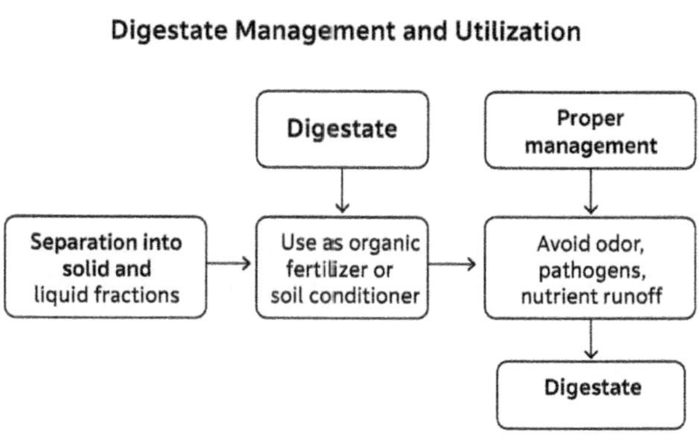

Fig. 3.2 Process of digestate management and utilization.

3.4.2 Use as organic fertilizer or soil conditioner

Digestate can be effectively used as an organic fertilizer or soil conditioner, offering both immediate and long-term benefits for agricultural productivity. When applied to fields, it supplies readily available nutrients such as nitrogen, phosphorus, and potassium, which support healthy crop growth and reduce dependence on synthetic fertilizers. The organic matter in digestate improves soil structure, enhances water retention, and promotes beneficial microbial activity, thereby increasing soil fertility over time. Solid digestate can be spread directly on farmland or composted to further stabilize its organic content, while the liquid fraction can be applied through irrigation systems for targeted nutrient delivery. By recycling the nutrients from agricultural residues back into the soil, digestate use closes the nutrient loop and supports sustainable farming practices.

3.4.3 Proper management of digestate

Proper management of digestate is essential to maximize its agricultural benefits while minimizing potential environmental and health risks. If not handled correctly, digestate can produce strong odors due to residual volatile compounds, creating nuisance issues for nearby communities. It may also contain pathogens, particularly if the digestion process has not operated at sufficiently high temperatures to achieve full sanitization. Furthermore, excessive or poorly timed application can lead to nutrient runoff into surface and groundwater, contributing to problems such as eutrophication. To prevent these issues, digestate should be stored in covered or enclosed facilities to reduce odor emissions and volatilization losses, applied to fields at agronomically appropriate rates, and incorporated into the soil promptly after spreading. Compliance with environmental regulations and the use of pathogen reduction techniques—such as pasteurization or extended storage—further ensure that digestate application remains safe, sustainable, and beneficial to agricultural systems.

3.4.4 Separation into solid and liquid fractions

Digestate can be mechanically separated into solid and liquid fractions to improve handling, storage, and utilization efficiency. The solid fraction, rich in organic matter, is well-suited for use as compost, soil amendment, or even livestock bedding after proper drying and stabilization. Composting the solid portion can further reduce pathogens, stabilize nutrients, and produce a more marketable organic fertilizer product. The liquid fraction, on the other hand, contains a high concentration of readily available nutrients such as ammonium nitrogen and potassium, making it ideal for fertigation through irrigation systems or as a bio-slurry applied directly to crops. This separation approach allows for targeted nutrient management, reduces transportation costs, and increases the versatility of digestate use in different farming systems, ultimately enhancing its value within a circular agricultural economy.

3.5 Summary

Biogas production from agricultural residues offers a sustainable approach to waste management, renewable energy generation, and nutrient recycling. Through the anaerobic digestion process, organic wastes are efficiently converted into biogas—a clean, renewable fuel—and digestate, a nutrient-rich byproduct that can enhance soil fertility. By adopting proper feedstock preparation, optimizing digestion conditions, and ensuring responsible digestate management, farmers and industries can maximize energy yields while reducing environmental impacts. This not only contributes to energy self-sufficiency and greenhouse gas mitigation but also promotes a circular economy in the agricultural sector.

References

Abanades, S., Abbaspour, H., Ahmadi, A., Das, B., Ehyaei, M.A., Esmaeilion, F., Assad, M.E.H., Hajilounezhad, T., Jamali, D.H., Hmida, A., Ozgoli, H.A., Safari, S., AlShabi, M., Bani-Hani, E.H., 2022. A critical review of biogas production and usage with legislations framework across the globe. Int. J. Environ. Sci. Technol. 1735-1472. 19 (4), 3377–3400. doi:10.1007/s13762-021-03301-6.

Abd, A.A., Othman, M.R., Shabbani, H.J.K., Helwani, Z., 2022. Biomethane upgrading to transportation fuel quality using spent coffee for carbon dioxide capture in pressure swing adsorption. J. Environ. Chem. Eng. 22133437. 10 (2). Available from: http://www.journals.elsevier.com/journal-of-environmental-chemical-engineering/. doi:10.1016/j.jece.2022.107169.

Aboudi, K., Fdez-Güelfo, L.A., Álvarez-Gallego, C.J., Romero-García, L.I., García-Morales, J.L., 2019. Pretreatments of digestates from anaerobic digestion for agronomic uses. In: Post Treatments of Anaerobically Treated Effluents. IWA Publishing, Spain, 263–282. Available from https://iwaponline.com/ebooks/book/750/Post-Treatments-of-Anaerobically-Treated-Effluents. doi:10.2166/9781780409740_0263.

Afridi, Z.U.R., Ullah, K., Mustafa, M.F., Saleem, H., Shaker, B., Ashraf, N., Aslam, S., 2023. Biogas as sustainable approach for social uplift in South East Asian Region. Energy Rep. 23524847. 10, 4808–4818. Available from: https://www.sciencedirect.com/science/journal/23524847. doi:10.1016/j.egyr.2023.11.037.

Ai, P., Jin, K., Alengebawy, A., Elsayed, M., Meng, L., Chen, M., Ran, Y., 2020. Effect of application of different biogas fertilizer on eggplant production: analysis of fertilizer value and risk assessment. Environ. Technol. Innov. 23521864. 19, 101019. doi:10.1016/j.eti.2020.101019.

Akbulut, A., Arslan, O., Arat, H., Erbaş, O., 2021. Important aspects for the planning of biogas energy plants: malatya case study. Case Stud. Therm. Eng. 2214157X. 26, 101076. doi:10.1016/j.csite.2021.101076.

Alengebawy, A., Mohamed, B.A., Jin, K., Liu, T., Ghimire, N., Samer, M., Ai, P., 2022a. A comparative life cycle assessment of biofertilizer production towards sustainable utilization of anaerobic digestate. Sustain. Prod. Consum. 23525509. 33, 875–889. Available from: http://www.journals.elsevier.com/sustainable-production-and-consumption/. doi:10.1016/j.spc.2022.08.016.

Alengebawy, A., Ran, Y., Ghimire, N., Osman, A.I., Ai, P., 2023. Rice straw for energy and value-added products in China: a review. Environ. Chem. Lett. 16103661. 21 (5), 2729–2760. Available from: https://www.springer.com/journal/10311. doi:10.1007/s10311-023-01612-3.

Alengebawy, A., Mohamed, B.A., Ran, Y., Yang, Y., Pezzuolo, A., Samer, M., Ping, A., 2022b. A comparative environmental life cycle assessment of rice straw-based bioenergy projects in China. Environ. Res. 00139351. 212 (Pt D), 113404. doi:10.1016/j.envres.2022.113404.

Alengebawy, A., Mohamed, B.A., Ghimire, N., Jin, K., Liu, T., Samer, M., Ping, A., 2022c. Understanding the environmental impacts of biogas utilization for energy production through life cycle assessment: an action towards reducing emissions. Environ. Res. 00139351. 213, 113632. doi:10.1016/j.envres.2022.113632.

Archana, K., Visckram, A.S., Kumar, P.S., Manikandan, S., Saravanan, A., Natrayan, L., 2024. A review on recent technological breakthroughs in anaerobic digestion of organic biowaste for biogas generation: challenges towards sustainable development goals. Fuel. 00162361. 358, 130298. doi:10.1016/j.fuel.2023.130298.

Ardolino, F., Arena, U., 2019. Biowaste-to-biomethane: an LCA study on biogas and syngas roads. Waste Manag. 0956053X. 87, 441–453. doi:10.1016/j.wasman.2019.02.030.

Arfan, M., Eriksson, O., Wang, Z., Soam, S., 2023. Life cycle assessment and life cycle costing of hydrogen production from biowaste and biomass in Sweden. Energy Convers. Manag. 01968904. 291, 117262. doi:10.1016/j.enconman.2023.117262.

Aryal, N., Ghimire, N., Bajracharya, S., 2020. Coupling of microbial electrosynthesis with anaerobic digestion for waste valorization. Adv. Bioenergy 5, 101–127. Available from https://www.sciencedirect.com/bookseries/advances-in-bioenergy. doi:10.1016/bs.aibe.2020.04.003.

Atelge, M.R., Krisa, D., Kumar, G., Eskicioglu, C., Nguyen, D.D., Chang, S.W., Atabani, A.E., Al-Muhtaseb, A.H., Unalan, S., 2020. Biogas production from organic waste: recent progress and perspectives. Waste Biomass Valoriz. 1877265X. 11 (3), 1019–1040. Available from: http://www.springer.com/engineering/journal/12649. doi:10.1007/s12649-018-00546-0.

CHAPTER 4

Bioethanol production from agricultural residues

I. Abernaebenezer Selvakumari[a], Kiruthika Shanmugam[a], J. Jayamuthunagai[a], B. Bharathiraja[b], and R. Praveen Kumar[c]
[a]Centre for Biotechnology, Alagappa College of Technology, Anna University, Chennai, Tamil Nadu, India
[b]Department of Biotechnology, Easwari Engineering College, Chennai, Tamil Nadu, India
[c]Department of Biotechnology, Arunai Engineering College, Chennai, Tamil Nadu, India

4.1 Introduction

The world is coping with the dual challenges of meeting rising energy demands and mitigating the environmental impacts of fossil fuel consumption, which currently accounts for about 78% of global energy production and contributes to severe air pollution, acid rain, and climate changes (Wang and Azam, 2024). The transportation sector is a significant contributor to this problem, with oil consumption estimated to rise from 84 million barrels per day to 116 million barrels by 2030 (Abbasi et al., 2021). However, the finite nature of fossil fuels and their devastating environmental impacts have sparked a growing interest in alternative fuels, particularly biomass-based biofuels, which offer a promising solution to reduce dependence on oil, lower production costs, and minimize environmental impact.

Waste valorization is a promising solution for the dual purpose of waste management and renewable energy production. It involves converting waste biomass into useful products such as biofuels, chemicals, and other valuable materials. This approach relies on the assumption that waste biomass still contains untapped polymeric substances that can be converted into energy or other chemical forms, making waste a valuable resource (Bakari et al., 2024). Waste-to-energy (WtE) technologies, such as thermochemical and biological processes, can be used to generate renewable energy, including heat, biogas, syngas, and bioethanol. These technologies provide cheap sources of energy that are crucial for industrial processes. Valorization research has evolved over the years, with many techniques and developments achieved in recent decades. Waste biomass, including lignocellulosic feedstocks, can be explored as sources of chemicals and biofuels. Various studies have demonstrated the potential of waste valorization, including the production of biogas

Bioenergy from Agricultural Residues
https://doi.org/10.1016/B978-0-443-24726-2.00007-9

through anaerobic digestion, gasification, and pyrolysis (Velvizhi et al., 2023; Thuppahige et al., 2025; Arun et al., 2019). These approaches can divert waste from landfills, reduce greenhouse gas (GHG) emissions, and generate electricity or heat. In addition, waste valorization can produce high-value chemicals, such as bioethanol and nanobioplastics, and offer additional advantages, including amelioration of waste mal-odors and environmental pollution, and reduction of waste volume. Overall, waste valorization is a much-needed technology for cost-effective and sustainable waste management options and generation of renewable energy.

In this regard, biomass, a renewable source, can be converted into eco-friendly energy using waste products such as agricultural residues, industrial waste, and municipal solid waste, providing a sustainable alternative to fossil fuels. Furthermore, the development of biorefineries and processing systems tailored to India's unique geographical range can enable the efficient production of biobased products, helping the country become self-dependent and socially stable while reducing its reliance on fossil fuels and promoting environmental sustainability (Raman et al., 2025). By harnessing the potential of biomass and other renewable energy sources, we can work towards a more sustainable energy future and mitigate the environmental impacts of fossil fuel consumption.

Biofuels are a promising alternative to fossil fuels, derived from biological resources such as plants and organic waste, offering a renewable energy source with a long history of use. They don't contribute to carbon emissions such as petrol or coal, making them a potential solution for reducing GHG emissions. Bioethanol is produced through the fermentation of starches or sugars from crops such as corn, sugarcane, and wheat and can be blended with gasoline to reduce emissions (Kazmi et al., 2025). Biodiesel, on the other hand, is derived from vegetable oils or animal fats through transesterification and can be used as a direct replacement for diesel fuel (Salaheldeen et al., 2021). Gaseous biofuels such as biogas and syngas are generated through anaerobic digestion and incomplete combustion of biomass, respectively, and can be used to generate electricity or heat (Enebe et al., 2023). Biobutanol, green diesel, and biohydrogen are also emerging as potential alternatives, with benefits including higher energy density, compatibility with existing engines, and near-zero emissions (Serrano-Echeverry et al., 2025). Biofuels can be further categorized into primary and secondary biofuels, with primary biofuels being raw materials used in their natural state and secondary biofuels being processed biofuels that can be used for a broader spectrum of applications. Overall, biofuels have the potential to play a significant role in mitigating climate change and improving energy sustainability, and their development and use are expected to continue growing in the coming years (Enebe et al., 2023).

Biofuels are categorized into four generations based on feedstock resources and generation technologies. First-generation (1G) biofuels, such as ethanol and biodiesel, are produced from edible biomass but face criticism for competing with the food supply. Second-generation (2G) biofuels utilize nonedible lignocellulosic biomass, such as agricultural residues and energy crops, offering a more sustainable alternative. Third-generation biofuels are derived from algae, which have high lipid content and can be grown on nonarable land, reducing competition with food crops. Fourth-generation biofuels involve capturing CO_2 and storing it or utilizing it through photobiological solar fuels and electrofuels, potentially revolutionizing biofuel production with a carbon-negative solution. Overall, biofuels have the potential to play a significant role in mitigating climate change and improving energy sustainability, with ongoing research aiming to improve efficiency, sustainability, and cost-effectiveness (Kazmi et al., 2025; Salaheldeen et al., 2021; Enebe et al., 2023; Serrano-Echeverry et al., 2025; Malik et al., 2024).

4.2 Bio-resources for bioethanol

Globally, biofuel production has been a growing industry, despite being negatively impacted by the pandemic. The primary biofuels are ethanol and biodiesel, which can substitute gasoline and diesel, significant contributors to GHGs and particulate matter. Most biofuels are currently produced from food crops such as maize, sugarcane, and vegetable oils, which can compete with food production (Ballesteros et al., 2004; El-Araby, 2024). Thus, agriculture is a significant biological sector, accounting for a substantial portion of global biomass production. With approximately 36% of the global land area suitable for agricultural production, there is a vast potential for biomass generation (Saleem, 2022). Agricultural residues, such as crop straws and pulses, offer a promising source for biofuel production, contributing to a more sustainable and renewable energy future.

The utilization of agricultural residues as a substrate for biofuel production is a highly promising approach that addresses multiple challenges simultaneously. India, being one of the world's largest producers of agricultural goods, generates an enormous amount of agrowaste, estimated to be around 500–550 million tons annually (Pratap Singh and Prabha, 2018).

Instead of letting this waste go to waste or being burned in open fields, contributing to environmental pollution and health hazards, agrowaste can be harnessed as a valuable resource for biofuel production. By leveraging various conversion technologies, such as anaerobic digestion, gasification, and fermentation, agrowaste can be transformed into biofuels such as biogas, bioethanol, and biochar. This not only provides a sustainable solution for waste management but also generates renewable energy, reducing dependence on fossil fuels and mitigating climate change. The economic worth of agricultural residues is substantial, estimated to be over $32 billion annually, and by leveraging waste-to-value strategies, farmers can reduce their reliance on expensive chemical fertilizers (Polipalli et al., 2025).

Agricultural residues such as wheat straw and corn stover are promising feedstocks for biofuel production, as they do not compete with food production. Bioconversion of these residues into bioproducts such as bioethanol, biohydrogen, and butanol has shown practical success, but challenges remain in pretreatment, enzyme lysis, and fermentation. By adopting sustainable practices and technologies, India can unlock the potential of agrowaste as a biofuel substrate, contributing to a more environmentally friendly and energy-secure future (Serrano-Echeverry et al., 2025; El-Araby, 2024). Pulses, belonging to the Leguminosae family, are a vital crop due to their high protein content and importance in human diets. They require minimal water and inputs, help maintain soil fertility through biological nitrogen fixation, and can mitigate GHG emissions. Pulse crop residues, rich in cellulose, hemicellulose, and lignin, can be converted into biofuels, providing a supplementary revenue stream for farmers and creating employment opportunities (Bhuvaneshwari et al., 2019). Specific crops such as black gram, pigeon pea, and pea have potential for biofuel production from their waste biomass. For instance, pigeon pea wood contains cellulose, hemicellulose, and lignin, making it a potential lignocellulosic resource for renewable energy (Tanquilut et al., 2019). Similarly, waste pea pods can be converted into biobutanol using modern technologies (Serrano-Echeverry et al., 2025).

Agricultural residues such as soybean straw and faba bean straw are rich in cellulose and hemicellulose, making them suitable for biofuel production (Akshaya et al., 2023). Legumes, which account for 44% of food waste, offer a significant opportunity for biofuel generation without competing with food production (Rațu et al., 2023). These crops can be grown on degraded lands or marginal soils, reducing GHG emissions and increasing carbon sequestration. Other biomass residues, such as peanut shells, can be used as a low-moisture energy source for heat production in industrial boilers and households (Saleem, 2022). Lignocellulosic biomass has been explored for biofuel generation since the late 20th century, offering a renewable, inexpensive, and plentiful source of energy. Energy crops such as maize, sugarcane, soybean, and sunflower can be converted into biofuels such as ethanol (Serrano-Echeverry et al., 2025). However, these crops often compete with food production. Alternatively, agricultural residues such as wheat straw, corn stover, and cereal straws can be utilized for biofuel production without impacting food security.

Globally, the remarkable harvest figures—such as Tajikistan's 9874 kg/ha of rice, New Zealand's 9668 kg/ha of wheat, Belgium's 8059 kg/ha of barley, Peru's 121,503 kg/ha of sugarcane, Tonga's 32,479 kg/ha of coconut, the Syrian Arab Republic's 88,800 kg/ha of bananas, and Israel's 50,094 kg/ha of groundnuts—mean that each hectare produces vast amounts of agricultural residues (Fig. 4.1). These leftover stalks, husks, bagasse, and shells represent a huge, untapped feedstock that can be collected and converted into biofuels, turning surplus crop waste into a valuable energy resource. Each hectare of high-yield rice generates roughly 12 t of straw (about 4200 L ethanol), wheat yields around 12.6 t of straw (≈4400 L ethanol), barley supplies ~9.7 t of stalks (≈3400 L ethanol), sugarcane produces up to 36 t of bagasse (≈9000 L ethanol or 70 MWh electricity), coconut husks

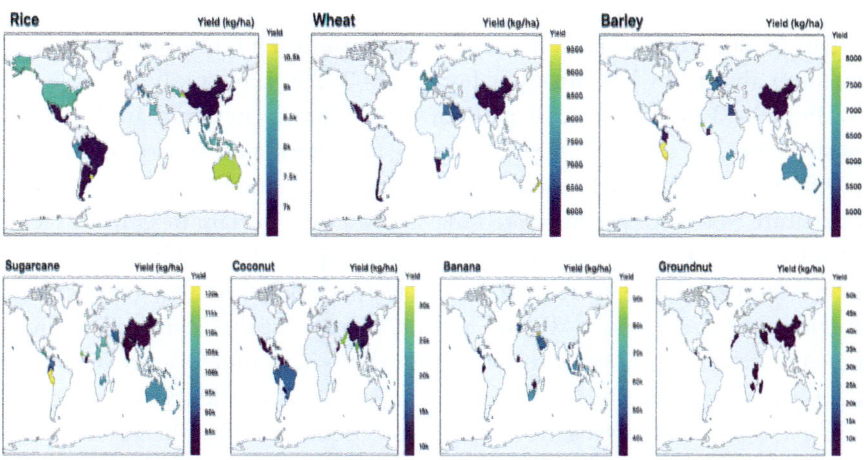

Fig. 4.1 Global agricultural residue potential. *Food and Agriculture Organization of the United Nations. (n.d.). Agricultural production statistics 2010-2023. Retrieved from: https://www.fao.org/statistics/highlights-archive/highlights-detail/agricultural-production-statistics-2010-2023/en.*

reach 19.5 t (≈5000 L ethanol), banana pseudostems and leaves amount to 71 t (≈18,000 L ethanol), and groundnut shells total 25 t (≈6000 L ethanol) (Food and Agriculture Organization of the United Nations, n.d.). By capturing and converting these lignocellulosic wastes through biochemical fermentation, thermochemical processes, or anaerobic digestion, the same high-productivity agriculture that feeds billions can also supply a substantial, sustainable feedstock for biofuels, reducing waste, lowering emissions, and creating new rural energy markets.

Across India's most productive states Fig. 4.2, the latest yield figures reveal a massive, yet largely untapped, reservoir of agricultural residues that could be harnessed for bio-fuel production: wheat reaches 5.00 t/ha in Chandigarh, 4.71 t/ha in Punjab and 4.66 t/ha in Haryana, generating roughly 6.5 t of straw per hectare (1.3 kg straw kg^{-1} grain)—enough to yield about 2300 L of ethanol; rice peaks at 10.70 t/ha in Goa, 5.5 t/ha in Chandigarh and 4.34 t/ha in Punjab, producing around 12.8 t of rice straw per hectare in Goa (≈ 4, 500 L ethanol) and still over 3 t of straw in lower-yield states such as Uttar Pradesh (2.68 t/ha) (≈ 1100 L ethanol); maize leads in Andhra Pradesh (7.14 t/ha), Tamil Nadu (7.01 t/ha) and Bihar (6.46 t/ha), delivering about 7 t of stalks and cobs per hectare (≈ 2, 500 L ethanol); groundnut yields are highest in West Bengal (3.06 t/ha) and Puducherry (2.95 t/ha), supplying roughly 1.5 t of shells and vines per hectare, suitable for thermochemical conversion to about 0.9 t of bio-oil; coconut production tops 14.66 t/ha in Andhra Pradesh and 11.51 t/ha in Tamil Nadu, with husks and fronds accounting for ~60 % of the biomass, yielding about 8.8 t of husk per hectare (≈ 2, 200 L ethanol); barley is led by Punjab (3.84 t/ha) and Rajasthan (3.40 t/ha), generating around 4.6 t of straw per hectare (≈ 1, 600 L ethanol); and banana dominates in Gujarat (79.89 t/ha) and Uttar Pradesh (54.28 t/ha), where pseudo-stems and leaves constitute ~80 % of the harvested mass, meaning a hectare in Gujarat yields about 64 t of wet biomass, which after drying provides roughly 12 t of dry matter and could be converted to about 4, 200 L of ethanol (Ministry of Agriculture and Farmers Welfare, n.d.). Collectively, these high-yield crops generate tens of millions of tons of lignocellulosic residues each year, and by deploying modest collection infrastructure, on-site shredders for maize and banana waste, and aggregation points for coconut husks, the residues can be channeled into biochemical, thermochemical, or anaerobic-digestion pathways, turning agricultural surplus into a substantial, low-carbon biofuel feedstock while reducing field-burning and creating new revenue streams for farmers.

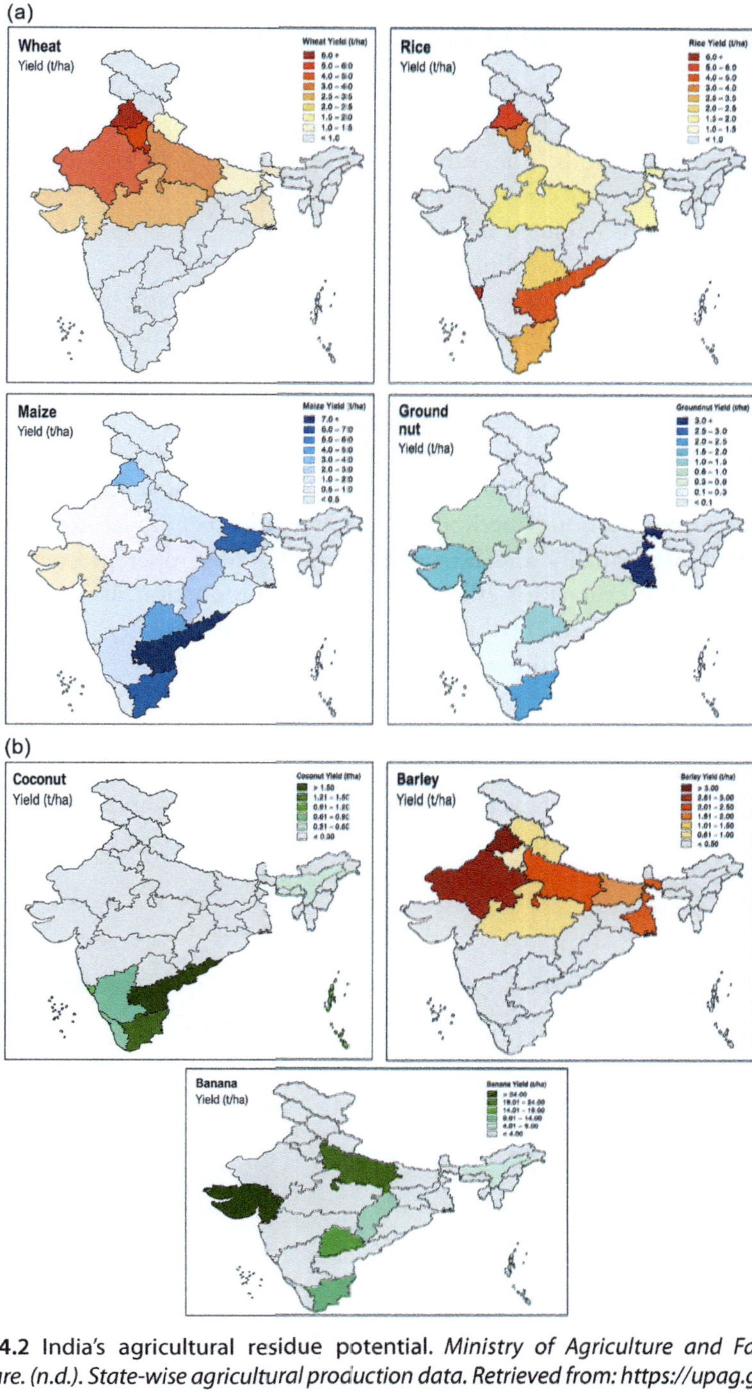

Fig. 4.2 India's agricultural residue potential. *Ministry of Agriculture and Farmers Welfare. (n.d.). State-wise agricultural production data. Retrieved from: https://upag.gov.in/dashreports/statewiseapy?rtab=Area%2C+Production+%26+Yield&rtype=reports.*

4.3 Bioconversion of biomass to bioethanol

Bioethanol production typically involves several key steps. First, the carbohydrate content in the substrate is prepared and broken down using pretreatment methods. The substrate is then converted into bioethanol through different processes (Kazmi et al., 2025).

The process is depicted in Fig. 4.3 as follows:

- Pretreating the lignocellulosic material
- Breaking down cellulose into sugars through enzymatic hydrolysis
- Converting these sugars into bioethanol through fermentation
- Recovering the bioethanol through distillation

The quality of bioethanol production starts with the preparation of raw materials. After harvesting crops, grains are separated, and the remaining plant parts can be utilized for bioethanol production. Factors such as soil processing, planting methods, and cultivation techniques significantly impact biomass quality and bioethanol production. Proper preparation of raw materials is crucial for efficient bioethanol production, as it affects the quality and yield of the final product (Velvizhi et al., 2023).

Different crops require specific preparation methods:

Faba bean: Leftover pods and straw are milled and dried before pretreatment.

Soybean: Soybean hulls are ground into smaller particles, and pretreatment methods such as thermomechanical extrusion and acid or alkali hydrolysis are optimized for bioethanol production.

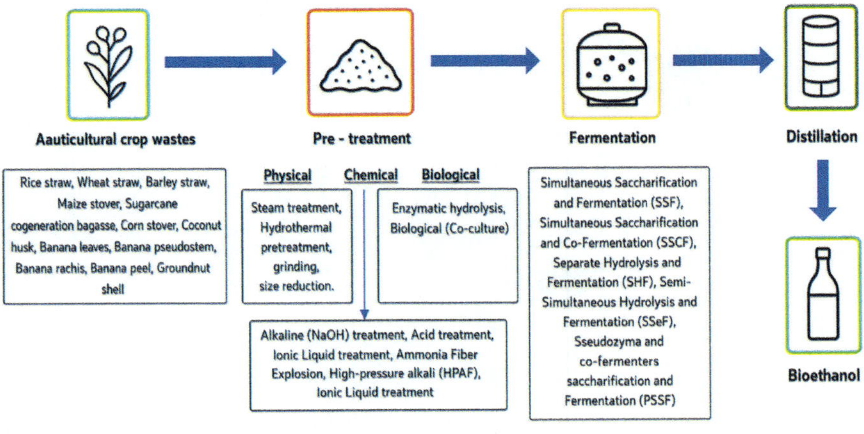

Fig. 4.3 Bioethanol production from agricultural wastes.

Pea and broad beans: Pea peels and pods are dried and homogenized to facilitate pretreatment and sugar release.

Pigeon pea: Stalks are dried and pulverized into fine particles to enable effective saccharification.

4.3.1 Lignocellulosic pretreatment

Pretreatment is a crucial step in converting lignocellulosic materials into bioethanol. It involves breaking down the complex matrix of cellulose, hemicellulose, and lignin to make the process more accessible to chemical or biological treatments. The goal is to reduce cellulose crystallinity, increase biomass surface area, and facilitate enzymatic action, ultimately leading to increased bioethanol production. There are four main types of pretreatment techniques: physical, chemical, physicochemical, and biological, often used in combination to achieve optimal results. These methods use approaches such as microbial, alkaline, acidic, steam, or thermal treatments to enhance bioethanol productivity. However, harsh conditions can generate inhibitors, affecting fermentation rates and yields. Pretreatment can account for up to 33% of total bioethanol production costs, making optimization essential for maximizing yields while minimizing energy input and costs (Damayanti et al., 2021). The choice of pretreatment method often depends on the raw material and specific bioethanol production goals. Effective pretreatment can increase the yield of monomeric sugars, reduce energy demands, and minimize costs, making it a vital step in the bioethanol production process (Table 4.1).

4.3.1.1 Physical pretreatment

Physical pretreatment methods aim to reduce the size and crystallinity of biomass, making it more accessible to downstream processing.

4.3.1.2 Mechanical size reduction

Mechanical size reduction involves comminution through milling, grinding, or chipping to reduce cellulose crystallinity and improve the efficiency of subsequent processing. This pretreatment method increases the surface area and reduces the crystallinity of biomass by decreasing particle size. These methods don't generate toxins; they can be energy-intensive. The choice of mill types and particle size depends on the biomass material and desired outcome, as very fine particle sizes can have negative effects such as increased energy consumption and clumping. Examples of successful applications include milling dried pigeon pea stalks and faba bean waste to boost saccharification yields (Duque et al., 2021).

Table 4.1 Effects of Raw Material Processing Techniques.

Characteristic	Process	Effect	Advantages	Disadvantages	Reference
Mechanical size reduction	Comminution	Increases surface area	No toxins	Energy-intensive	Duque et al. (2021)
Pyrolysis	Thermal decomposition	Converts to bio-oil	Low-ash bio-oil	High water content	Hasan and Avami (2021)
Microwave/ Electron beam	Thermal/nonthermal disruption	Improves disruption	Improves biomass conversion	–	Han et al. (2020)
Advanced oxidation	Steaming	Enhances bioethanol yields	Effective	–	Panahi et al. (2022)
Steam explosion	High-pressure steam	Increases cellulase accessibility	High xylose yields	–	Thuppahige et al. (2025)
Liquid hot water	Compressed hot water	High xylose recovery	High xylose recovery	Produces toxic compounds	Konti and Damigos (2018)
AFEX	Alkaline thermal	Exposes lignocellulosic materials	No downstream inhibitors	Requires ammonia recovery	Luo et al. (2010)
CO_2 explosion	High-pressure steam	Higher conversion yields	Cost-effective	–	Damayanti et al. (2021)
Acid	Acid hydrolysis	Improves cellulose accessibility	Increases glucose yields	Produces inhibitors	Rațu et al. (2023)
Alkaline	Alkaline digestion	Increases enzyme accessibility	Decreases cellulose crystallinity	–	Phwan et al. (2019)
Organosolv	Organic solvent delignification	Delignifies biomass	Eliminates liquor burning	–	Han et al. (2020)
Extrusion	Thermo-mechanical	Frees cellulose microfibrils	Continuous operation	–	Ramanathan et al. (2022)
Wet oxidation	High temperature and pressure	Generates pentose monomers	Improves hemicellulose degradation	Requires careful control	Mankar et al. (2021)
Biological	Microbial degradation	Liberates cellulose	Safe and energy-saving	Low hydrolysis rates	Miret et al. (2016)

4.3.1.3 Pyrolysis

Pyrolysis is an endothermic process that decomposes biomass at high temperatures, producing gaseous products and residual char. The residual char can be further treated with water or mild acid to produce a carbon source for microbial growth. It is a thermochemical process that converts lignocellulose into bio-oil. For example, pigeon pea wood was processed in a semi-continuous reactor at 466°C with a controlled heating rate and nitrogen flow, yielding 52% bio-oil with low ash content and high heating value (Hasan and Avami, 2021). However, the high-water content in bio-oil can limit its utility, making further processing necessary.

4.3.1.4 Microwave oven and electron beam irradiation pretreatment

Microwave pretreatment uses thermal and nonthermal effects to disrupt the recalcitrant structures of lignocellulose. This method can improve the disruption of biomass and increase the yield of reducing sugars. Electron beam irradiation is a nonthermal method that uses high-energy radiation to alter the chemical and physical properties of biomass, increasing the specific surface area and decreasing the degree of polymerization and crystallinity of cellulose. Both methods have shown promise in improving the efficiency of biomass conversion to bioethanol (Han et al., 2020).

4.3.1.5 Physiochemical pretreatment

Physicochemical pretreatment methods use a combination of physical and chemical processes to break down biomass and make it more accessible to enzymatic attack, ultimately leading to increased bioethanol production.

4.3.1.6 Advanced oxidation processes

Advanced oxidation processes, such as steaming pretreatment, are effective methods that involve treating biomass with steam, wet air, high temperature, or high pressure to enhance bioethanol yields. For example, steaming pretreatment of pigeon pea stalks involves mixing dried solids with water and heating to 121°C for 30 min, yielding significant amounts of reducing sugars and pentose (Panahi et al., 2022).

4.3.1.7 Steam explosion or autohydrolysis

Steam explosion is a promising method that uses high-pressure steam to fractionate biomass and make it more accessible to cellulase attack. This method can yield high amounts of xylose and other valuable compounds (Thuppahige et al., 2025).

4.3.1.8 Liquid hot water method

The liquid hot water method uses compressed hot water to hydrolyze hemicellulose and release oligomers. While this method can achieve high xylose recovery rates, it can also produce toxic compounds that inhibit microbial growth (Konti and Damigos, 2018).

4.3.1.9 Ammonia fiber explosion (AFEX)

Ammonia fiber explosion (AFEX) is an alkaline thermal pretreatment that uses liquid ammonia and steam explosion to expose lignocellulosic materials. This method is effective for biomass with low lignin content and does not produce inhibitors of downstream processes. However, it requires efficient ammonia recovery to be economical (Luo et al., 2010).

4.3.1.10 CO_2 explosion

CO_2 explosion is a cost-effective method that acts similarly to steam and ammonia explosion techniques. It does not cause the formation of inhibitors and can achieve higher conversion yields compared to steam explosion (Damayanti et al., 2021).

4.3.1.11 Chemical pretreatment

Chemical pretreatment methods involve using chemicals to break down biomass and make it more accessible to enzymatic attack.

4.3.1.12 Acid pretreatment

Acid pretreatment uses dilute or concentrated acids to hydrolyze hemicellulose and improve cellulose accessibility. However, acid pretreatment can produce inhibitors such as furfural and 5-hydroxymethylfurfural, which can affect microbial growth (Rațu et al., 2023).

4.3.1.13 Alkaline pretreatment

Alkaline pretreatment uses alkaline solutions such as sodium hydroxide to digest the lignin matrix and make cellulose and hemicellulose available for enzymatic degradation. This method can decrease cellulose crystallinity and increase the accessibility of biomass to enzymes (Phwan et al., 2019).

4.3.1.14 Organosolv pretreatment

Organosolv pretreatment uses organic solvents such as methanol, ethanol, or acetic acid to delignify biomass and make it more accessible to enzymatic attack. This method can eliminate the need to burn liquor and allow for the isolation of lignin (Han et al., 2020).

These methods can be effective in breaking down biomass and increasing bioethanol production. These methods are easy to operate and can achieve good conversion yields in a short time. Examples of successful applications include dilute acid and alkali pretreatments that increased glucose yields from soybean hulls, and similar treatments applied to other biomass materials such as faba bean, pigeon pea, pea, and broad bean pods.

4.3.1.15 Thermo-mechanical pretreatment

Extrusion is a thermomechanical pretreatment method that uses high pressure, temperature, and speed to break down cell wall structures and free cellulose microfibrils. This method is suitable for continuous operation, offers high product stability, and generates less effluent waste. Extrusion can be integrated with alkali or acid pretreatments to increase sugar recovery. For example, soybean waste was extruded using a twin-screw extruder with controlled temperature, screw speed, and moisture levels, achieving 95% cellulose to glucose conversion (Ramanathan et al., 2022).

4.3.1.16 Wet oxidation

Wet oxidation involves subjecting biomass to high temperature and pressure to hydrolyse and oxidize lignin, generating pentose monomers from hemicellulose. This method requires careful control of temperature, incubation time, and oxygen pressure. For example, faba bean straw was treated with wet oxidation at 195°C and 12 bar oxygen pressure, with sodium carbonate added to improve hemicellulose degradation and reduce inhibitor production (Mankar et al., 2021). The pretreated materials can be stored or processed further for bioethanol production.

4.3.1.17 Biological pretreatment

Biological pretreatment uses microorganisms such as fungi to break down the lignocellulosic complex and liberate cellulose. White rot fungi are particularly effective in degrading lignin and hemicellulose. This method is safe, energy-saving, and does not require chemicals. However, it has limitations, including low hydrolysis rates and yields, which can impede its implementation. Biological pretreatment can be effective in certain conditions, such as low temperatures, and can be enhanced with nutrient supplements. While it shows promise, the long processing times required for bio-delignification remain a challenge (Miret et al., 2016).

4.3.2 Enzymatic hydrolysis

Enzymatic hydrolysis is a critical step in bioethanol production where complex carbohydrates are converted to simple monomers using enzymes such as cellulase and hemicellulase. This process requires mild conditions and less energy compared to acid hydrolysis. Cellulase enzymes, such as endoglucanase, exoglucanase, and beta-glucosidase, break down cellulose into glucose, while hemicellulolytic enzymes break down hemicellulose into pentoses and hexoses. Microorganisms such as *Trichoderma* and *Aspergillus* are commonly used to produce these enzymes. Factors such as temperature, pH, substrate concentration, and enzyme loading affect the yields of monomer sugars (Rezania et al., 2020). Efficient pretreatment and surfactant addition can enhance enzymatic hydrolysis, leading to high sugar yields and improved bioethanol production.

4.3.3 Saccharification and fermentation

Saccharification, a critical step in bioethanol production, involves liberating reducing sugars from pretreated cellulosic biomass using enzymes such as cellulase and beta-glucosidase. These enzymes break down cellulose into fermentable sugars, making them available for fermentation. By optimizing saccharification and fermentation processes, bioethanol production can become more efficient and sustainable.

Fermentation is a critical step in bioethanol production, where microorganisms such as yeast or bacteria convert saccharified biomass into ethanol. However, native microorganisms often struggle to efficiently ferment both pentose and hexose sugars, limiting industrial utilization. To overcome this, genetically modified or engineered microorganisms are employed to achieve complete sugar utilization and improved production benefits (Thatoi et al., 2016). Various fermentation processes are used, including:

1. **Separate hydrolysis and fermentation:** A two-step process where saccharification and fermentation occur separately, allowing for optimal conditions for each step.
2. **Simultaneous saccharification and fermentation:** A single-step process where saccharification and fermentation occur simultaneously, improving yields and cost-effectiveness. However, temperature differences between enzyme hydrolysis and fermentation can pose limitations.
3. **Consolidated bioprocessing:** A process that combines enzyme production, saccharification, and fermentation in a single step, potentially reducing costs and increasing efficiency.

4. **Simultaneous saccharification and cofermentation:** A process that combines saccharification and cofermentation of multiple sugars, improving yields and efficiency.

Native and genetically modified microorganisms, such as *Saccharomyces cerevisiae*, a popular yeast strain renowned for its high ethanol tolerance and yield, *Escherichia coli,* a bacterium that can be engineered for enhanced bioethanol production, and *Zymomonas mobilis*, a bacterium known for its high ethanol tolerance and yield, are pivotal in bioethanol production. Genetic engineering has significantly contributed to developing these microorganisms with improved substrate utilization, yield, and recovery rates, enabling them to efficiently ferment both pentose and hexose sugars (Yusuf and Inambao, 2019). This advancement makes bioethanol production more efficient and cost-effective, paving the way for sustainable biofuel development.

4.3.3.1 Types of fermenters

In bioethanol fermentation, various types of fermenters are utilized to optimize the conversion of sugars into ethanol. These include batch fermenters, which are simple and suitable for small-scale production, and continuous stirred tank reactors, which offer high productivity and consistent product quality for large-scale operations. Plug flow reactors provide high productivity with minimal mixing, while tower fermenters, with their tall, vertical design, facilitate efficient mixing and aeration. Air-lift and bubble column fermenters use air bubbles for mixing and aeration, reducing mechanical stress on microorganisms. Fed-batch fermenters allow for controlled substrate feeding to improve yields, and vacuum fermenters enable simultaneous fermentation and product removal to enhance efficiency (Kazmi et al., 2025). Each fermenter type has its advantages and disadvantages, and the choice depends on factors such as scale, feedstock, microorganism, and process requirements, ultimately influencing the efficiency and cost-effectiveness of bioethanol production (Table 4.2).

4.3.4 Bioethanol separation

The production of bioethanol involves a series of complex processes, with product recovery and purification being crucial steps. Distillation, an energy-intensive process, is typically employed to separate bioethanol from the fermentation broth, leveraging differences in boiling points to yield high-purity ethanol (Han et al., 2020). However, achieving anhydrous ethanol is challenging due to the formation of an azeotrope

Table 4.2 Bioethanol production from agricultural residue.

Sl. No	Residue	Carbon source	Microorganism	Pre treatment	Method	Bioethanol yield	Reference
	Rice straw	Cellulose	*S. cerevisiae*	Enzymatic hydrolysis	Fermentation	0–11 g/L	Belal (2013)
		Cellulose	*S. cerevisiae*	Popping pretreatment, enzymatic hydrolysis	Fermentation	0.172 g/g	Wi et al. (2013)
		Cellulose	*S. cerevisiae*	Acid + Ultrasound	Fermentation	1.69% (v/v)	Yoswathana (2010)
	Wheat straw	Glucose and xylose	*S. cerevisiae*	Enzymatic hydrolysis	Fermentation	1.16 g/kg · h	Talebnia et al. (2010)
		Cellulose	*K.marxianus* MTCC 4136	Dilute sulfuric acid + hydrothermal sterilization	Fermentation	21.6 g/L	Singhania et al. (2014)
		Cellulose	*K. fragilis* CECT 10,875	Steam treatment	Fermentation	18.1 g/L	Ballesteros et al. (2004b)
	Barley straw	Cellulose	*S. cerevisiae*	Steam treatment	Fermentation	22.4 g/L	Ballesteros et al. (2004b)
	Sweet sorghum bagasse	Cellulose, hemicellulose, and lignin	*K.marxianus* CECT 10,875	Steam treatment	Fermentation	16.2 g/L	Öhgren et al. (2007)
	Corn Stover	Cellulose, hemicellulose	*S. cerevisiae*	Steam treatment	Simultaneous saccharification and fermentation	72.4%	Öhgren et al. (2006)
		Glucan, xylan	Recombinant*S. cerevisiae*	Steam treatment	Simultaneous saccharification and fermentation	59%	Liu and Chen (2016)
		Glucose, xylose	*S. cerevisiae IPE003*	Steam treatment	Simultaneous saccharification and cofermentation	>90%	Kazi et al. (2010)

		Cellulose, Hemicellulose	*Z. mobilis*	Acid treatment	Separate hydrolysis and fermentation	46%–50%	Mirmohamadsadeghi et al. (2016)
		Cellulose	*P. stipitis*	Alkaline treatment	Simultaneous Saccharification and Fermentation	65%–70%	Qing et al. (2017)
		Glucan	*S. cerevisiae*	Organo solvent treatment (Methanol + HCl catalyst)	Fermentation	98.6%	Kedong and Ruan (2015)
		Cellulose	*T. reesei* + *C. comatus*	Biological treatment	Simultaneous biological pretreatment + Hydrolysis	82%	Teymouri et al. (2005)
		Cellulose, hemicellulose	*S. cerevisiae*	Ammonia fiber explosion	Simultaneous saccharification and fermentation	~90%	Uppugundla et al. (2014)
		Cellulose, hemicellulose	*S. cerevisiae*	Ionic liquid treatment	Simultaneous saccharification and fermentation	85%–90%	Song et al. (2013)
		Cellulose	*I. lacteus*	Biological treatment + Mn^{2+} enhancement	Enzymatic hydrolysis	61.4%	Hernawan et al. (2017)
	Sugarcane bagasse	Cellulose, hemicellulose	*S. cerevisiae*	Alkaline pretreatment	Simultaneous saccharification and fermentation using cellulase	5.87 ± 0.78%	Hernández-Salas et al. (2009)
		Cellulose, hemicellulose	*S. cerevisiae*	Alkaline treatment	Simultaneous saccharification and fermentation using cellulase + β-glucosidase	6.83 ± 0.07%	

Continued

Table 4.2 Bioethanol production from agricultural residue—cont'd

Sl. No	Residue	Carbon source	Microorganism	Pre treatment	Method	Bioethanol yield	Reference
		Cellulose, hemicellulose	*S. cerevisiae*	Acid hydrolysis	Batch fermentation	14.11%	Cheng et al. (2008)
		Cellulose, hemicellulose	*S. cerevisiae*	Acid hydrolysis	Batch fermentation	15.72 %	Cheng et al. (2008)
		Cellulose, hemicellulose	*P. tannophilus* DW06	Acid hydrolysis + Electrodialysis	Batch fermentation	34%	Chandel et al. (2007)
		Cellulose, hemicellulose	*P. tannophilus* DW06	Acid hydrolysis	Batch fermentation	0.03%	Chandel et al. (2007)
		Cellulose, hemicellulose	*C. shehatae* NCIM 3501	Ion exchange resin (DIAION HPA 25), acid hydrolysis	Batch fermentation	8.67 g/L	MARTIN et al. (2007)
		Cellulose, hemicellulose	*C. shehatae* NCIM 3501	Activated charcoal, acid hydrolysis	Batch fermentation	7.43 g/L	
		Cellulose, hemicellulose	*C. shehatae* NCIM 3501	Enzymes hydrolysis (laccase from *C. stercoreus*), acid hydrolysis	Batch fermentation	6.50 g/L	
		Cellulose, hemicellulose	*C. shehatae* NCIM 3501	Overliming, acid hydrolysis	Batch fermentation	5.19 g/L	
		Cellulose, Hemicellulose	*C. shehatae* NCIM 3501	Neutralization, acid hydrolysis	Batch fermentation	3.46 g/L	
		Cellulose, hemicellulose	*S. cerevisiae*	Alkaline treatment + Enzymatic saccharification	Batch fermentation	32.57%	Cheng et al. (2008)

	Cellulose, hemicellulose	*S. cerevisiae*	Alkaline treatment + Enzymatic saccharification	Batch fermentation	25.76%	
	Cellulose, hemicellulose	Recombinant *S. cerevisiae*	H_2SO_4-catalyzed steam pretreatment	Batch fermentation	38%	Kaur et al. (2012)
	Cellulose, hemicellulose	Recombinant *S. cerevisiae*	H_2SO_4-catalyzed steam pretreatment	Batch fermentation	18%	
Cotton stalk	Glucan, xylan	*Pichia kudriavzevii HOP-1*	Alkaline treatment	Enzymatic hydrolysis + fermentation	19.82 g/L	Baig and Dharmadhikari (2014)
	Cellulose	*S. cerevisiae* + *P. tannophilus*	Alkaline treatment	Enzymatic hydrolysis + fermentation	9.56 g/L ethanol	Jiang et al. (2015)
	Cellulose, hemicellulose	–	Liquid hot water treatment	Hydrolysis + fermentation	0.18 g/g	Wang et al. (2016)
	Cellulose, hemicellulose	–	High-pressure alkali treatment	Enzymatic hydrolysis + fermentation	45.53%	Singh et al. (2017)
	Cellulose, hemicellulose	*S. cerevisiae*	$FeCl_3$ treatment	Enzymatic hydrolysis + fermentation	9.8 g/L	Ding et al. (2012)
Coconut husk	Cellulose, hemicellulose	*S. cerevisiae* ATCC 36,858	Microwave-assisted alkaline treatment	Simultaneous hydrolysis and fermentation	–	Vaithanomsat et al. (2011)
	Cellulose, hemicellulose	*S. cerevisiae*	Alkaline Treatment	Separate hydrolysis and fermentation, simultaneous hydrolysis and fermentation	≈ 85%	Gonçalves et al. (2014)

Continued

Table 4.2 Bioethanol production from agricultural residue—cont'd

Sl. No	Residue	Carbon source	Microorganism	Pre treatment	Method	Bioethanol yield	Reference
		Cellulose, hemicellulose	*S. cerevisiae*, *P. stipitis*, *Z. mobilis*	Alkaline H_2O_2 + alkaline delignification treatment	Simultaneous hydrolysis and fermentation	*S. cerevisiae*: 0.43 (g ethanol/g sugars); *P. stipitis*: 0.40 (g ethanol/g sugars); *Z. mobilis*: 0.42 (g ethanol/g sugars)	Gonçalves et al. (2015)
		Cellulose, hemicellulose	*S. cerevisiae*, *P. stipitis*, *Z. mobilis*	Alkaline H_2O_2 + alkaline delignification treatment	Semi-simultaneous hydrolysis and fermentation	*S. cerevisiae*: 0.45 (g ethanol/g sugars); *P. stipitis*: 0.43 (g ethanol/g sugars); *Z. mobilis*: 0.44 (g ethanol/g sugars)	
		Cellulose, hemicellulose	*S. cerevisiae*, *P. stipitis*, *Z. mobilis*	Autohydrolysis	Simultaneous hydrolysis and fermentation	*S. cerevisiae*: 0.44 (g ethanol/g sugars); *P. stipitis*: 0.43 (g ethanol/g sugars); *Z. mobilis*: 0.43 (g ethanol/g sugars)	Gonçalves et al. (2016)

	Cellulose, hemicellulose	*S. cerevisiae*, *P. stipitis*, *Z. mobilis*	Autohydrolysis	Semi-simultaneous hydrolysis and fermentation	*S. cerevisiae*: 0.45 (g ethanol/g sugars); *P. stipitis*: 0.44 (g ethanol/g sugars); *Z. mobilis*: 0.45 (g ethanol/g sugars)	
	Cellulose, hemicellulose	*S. cerevisiae*, *P. stipitis*, *Z. mobilis*	Hydrothermal catalyzed with NaOH	Simultaneous hydrolysis and fermentation	*S. cerevisiae*: 0.44 (g ethanol/g sugars); *P. stipitis*: 0.45 (g ethanol/g sugars); *Z. mobilis*: 0.43 (g ethanol/g sugars)	Soares et al. (2016)
	Cellulose, hemicellulose	*S. cerevisiae*, *P. stipitis*, *Z. mobilis*	Hydrothermal catalyzed with NaOH	Semi-simultaneous hydrolysis and fermentation	*S. cerevisiae*: 0.47 (g ethanol/g sugars); *P. stipitis*: 0.46(g ethanol/g sugars); *Z. mobilis*: 0.47(g ethanol/g sugars)	

Continued

Table 4.2 Bioethanol production from agricultural residue—cont'd

Sl. No	Residue	Carbon source	Microorganism	Pre treatment	Method	Bioethanol yield	Reference
		Cellulose, hemicellulose	*S. cerevisiae* (Ethanol Red, GSE16–T18)	Alkaline treatment	Separate hydrolysis and fermentation	0.43 (g ethanol/g sugars)	Ebrahimi et al. (2017)
		Cellulose, hemicellulose	*S. cerevisiae* (Hansen 2055)	Acidified aqueous glycerol	Simultaneous hydrolysis and fermentation	–	Soares et al. (2017)
		Cellulose, hemicellulose	*S. cerevisiae* (Hansen 2055)	Aqueous glycerol	Simultaneous hydrolysis and fermentation	–	
		Cellulose, hemicellulose	*S. cerevisiae* (GSE16–T18)	Alkaline treatment	Separate hydrolysis and fermentation	0.41 (g ethanol/g sugars)	Suhag et al. (2020)
	Banana leaves	Cellulose, hemicellulose	*S. cerevisiae*	Alkali–acid treatment	Fermentation	15.43 g/L	Shankar et al. (2020)
		Cellulose, hemicellulose	*S. cerevisiae*	Acid treatment	Fermentation	8.1 g/L	Harish et al. (2010)
		Cellulose, hemicellulose	*C. thermocellum*, *C. thermosaccharolyticum*	Alkali–acid treatment	Coculture fermentation	22 g/L	Gatdula et al. (2021)
	Banana pseudo-stem	Cellulose, hemicellulose	*K. marxianus*	Acid treatment	Simultaneous hydrolysis and fermentation	5.35 g/L	Meena et al. (2015)
		Cellulose, hemicellulose	*S. cerevisiae*	Biological treatment	Fermentation	9.8 g/L	Guerrero et al. (2018)
		Cellulose, hemicellulose	*S. cerevisiae*	Biological treatment	Fermentation	14.4 g/L	
	Banana rachis	Cellulose, hemicellulose	*S. cerevisiae*	Acid-catalyzed steam explosion	Prehydrolysis and simultaneous accharification and fermentation	48.3 g/L	Oberoi et al. (2011)

	Banana peel	Starch, hemicellulose	*S. cerevisiae*	Hydrothermal sterilization	Simultaneous saccharification and fermentation	28.2 g/L	Arumugam and Manikandan (2011)
		Hemicellulose	*S. cerevisiae*	Liquid hot water + dilute acid Alkaline treatment	Fermentation	13.84 g/L	Matharasi et al. (2018)
		Starch-based hydrolysate	*S. cerevisiae*	Mechanical pretreatment	Fermentation	17 g/L	Palacios et al. (2017)
		Cellulose, hemicellulose	*K. marxianus*	Acid treatment	Fermentation	21 g/L	Palacios et al. (2021)
		Cellulose, hemicellulose	*S. cerevisiae*	Autoclave treatment	Presaccharification and simultaneous saccharification and fermentation	32.6 g/L	Nyachaka et al. (2013)
	Groundnut shell	Cellulose, hemicellulose	*S. cerevisiae, A. niger*	Acid hydrolysis	Fermentation	55.8 ml	Akpan, et al. (2005)
		Cellulose, hemicellulose	*A. niger, S. cerevisiae*	Alkaline treatment	Fermentation	8.2%	Bhatt and Shilpa (2014)
		Lignocellulose	*B.stearothermophilus, S. cerevisiae*	Steam explosion + acid/alkali treatment	Simultaneous saccharification and fermentation	16.11%	Rabah et al. (2011)
		Cellulose, hemicellulose	*Z. mobilis, S. cerevisiae*	Enzymatic hydrolysis	Fermentation	0.955% (v/v)	Nwaze et al. (2023)
		Cellulose	*S. cerevisiae*	Acid hydrolysis	Fermentation	27.5%	Rabah et al. (2011)

(95.6% ethanol), necessitating additional dehydration steps such as molecular sieves or azeotropic distillation. To enhance efficiency and reduce costs, advanced separation technologies such as membrane-based methods and adsorption are being integrated into the process, contributing to energy savings and process intensification (Salaheldeen et al., 2021; Enebe et al., 2023).

4.4 Life cycle assessment

In the context of bioethanol production from agricultural residual biomass, conducting a life cycle assessment (LCA) that considers the transportation of biomass is crucial for gaining insights into the environmental implications of this stage in the bioethanol life cycle (Beagle and Belmont, 2019). To achieve optimal bioethanol production from such biomass, performing a sensitivity analysis on parameters including plant efficiency and the heating values of biomass is vital for understanding the influence of various variables on the overall performance of bioethanol production (Kumar and Verma, 2020). When evaluating the sustainability and environmental consequences of producing bioethanol from agricultural residual biomass, it is necessary to consider several key environmental factors. These encompass the quality of air and water, the profile of GHG emissions throughout the production process, the sequestration of carbon through agricultural practices, and the potential impacts on ecosystems due to eutrophication and atmospheric acidification (Caffrey and Veal, 2013). Various studies have investigated both the environmental and technical dimensions of different bioethanol production methods (Chen et al., 2020). For a thorough understanding of the environmental implications specific to each method of converting agricultural residual biomass into bioethanol, implementing a tailored LCA for each conversion process is requisite. This approach ensures a comprehensive evaluation of the environmental footprint associated with bioethanol production from agricultural residues.

4.5 Technoeconomic analysis

A technoeconomic analysis (TEA) for bioethanol production from agricultural residues evaluates the economic viability and technical feasibility of converting this biomass into bioethanol. In the context of India, where the government is promoting the blending of ethanol in transportation fuels, the production of 2G bioethanol from cellulosic biomass such as agricultural residues is being explored. Key findings from TEA studies indicate that while the type of feedstock has a minor impact on the production cost of bioethanol, economies of scale play a significant role (Gautam et al., 2026; Joseph et al., 2023). Furthermore, government support in the form of viability gap funding (VGF) can enhance the financial attractiveness of bioethanol production from agricultural residues (Oyegoke et al., 2021). However, if bioethanol is sold at prices comparable to those of 1G ethanol, VGF alone might not suffice to incentivize private investment adequately. The choice of process technologies, such as pervaporation for bioethanol dehydration, and the availability of abundant and inexpensive agricultural residues are critical factors that make bioethanol production promising (Zhao et al., 2015; Quintero et al., 2013). Considering these aspects, additional policy measures beyond VGF could be necessary for establishing a viable cellulosic bioethanol industry in India.

4.6 Challenges

Agro-wastes pose significant challenges for bioethanol production, primarily due to their seasonal availability, which makes ensuring a continuous supply difficult. The dispersed nature of agro-wastes increases collection and transportation costs, making it economically challenging. In addition, agro-wastes are prone to spoilage and degradation, requiring proper storage and preservation techniques to maintain their quality (Khounani et al., 2019). High moisture content in agro-wastes can lead to energy losses during conversion processes, reducing overall efficiency. The complex lignocellulosic structure of agro-wastes, comprising lignin, cellulose, and hemicellulose, necessitates pretreatment to release fermentable sugars, adding to processing costs. Furthermore, agro-wastes have alternative uses, such as animal feed, composting, or burning, which can impact their availability for bioenergy production. Establishing efficient supply chains and infrastructure is crucial, but logistical challenges and cost-effectiveness remain significant hurdles (Kumar et al., 2019). Addressing these challenges requires innovative solutions, including developing efficient collection and transportation systems, implementing cost-effective storage and preservation techniques, optimizing pretreatment processes, and exploring alternative conversion technologies (Smith et al., 2019). Policy support and investment in agro-waste-based bioenergy infrastructure are also essential to overcome these challenges and promote sustainable bioenergy production.

4.7 Policies and regulatory framework

The Indian government's initiatives to promote bioethanol production from agro-residues are comprehensive and multifaceted. The Pradhan Mantri JI-VAN Yojana, launched in 2019, supports the establishment of commercial-scale biorefineries utilizing lignocellulosic biomass and other renewable feedstock, providing financial assistance of up to 25% of the project cost or Rs. 150 crore per biorefinery (Press Information Bureau (PIB), Government of India, 2022). The National Biofuel Policy (2018) targets 20% ethanol blending in gasoline by 2025, encouraging the use of agricultural residues as feedstock, thereby reducing oil imports, promoting sustainable energy, and creating new revenue streams for farmers (Ministry of Petroleum and Natural Gas (MoPNG), Government of India, 2018). The Ethanol Blended Petrol Program mandates blending ethanol with gasoline, reducing fossil fuel dependence and emissions. In addition, the BioE3 Policy fosters biotechnology-driven manufacturing, including enzyme production for bioethanol, reducing reliance on imports (Press Information Bureau (PIB), Government of India; Ministry of Petroleum and Natural Gas (MoPNG), 2025). The government also offers financial incentives such as capital subsidies, tax benefits, and low-interest loans to promote bioethanol production. These initiatives aim to reduce India's dependence on fossil fuels, promote sustainable energy, and contribute to climate change mitigation, aligning with the country's goal of achieving carbon neutrality. By leveraging these policies and schemes, India can enhance energy security, boost the rural economy, and create a more sustainable future.

4.8 Future directions

Agro-residue-based bioethanol production requires a well-integrated system spanning from fields to final bioethanol production, considering key factors' impact on climate, habitat, and management. Integrated management of agro-residues, including collection, storage, and transportation, is crucial for ensuring a consistent and affordable feedstock supply, achievable through collaborations between farmers, landowners, and bioethanol producers. Understanding the microbiome's role in soil health and plant productivity informs strategies for optimizing agro-residue production and minimizing environmental impacts. Recognizing soil build-up and regulatory processes helps increase soil accrual and reduce nitrogen oxide emissions, with sustainable agricultural practices such as crop rotation and cover cropping enhancing soil health. Bioethanol production contributes to climate change mitigation by reducing GHG emissions and promoting carbon sequestration. Engaging policymakers, landowners, and farmers, emphasizing environmental benefits and opportunities, promotes sustainable bioenergy production, incentivizing sustainable practices, rural development, and reduced fossil fuel dependence. Effective integration of these factors achieves a well-functioning bioenergy system benefiting the environment and local communities, promoting sustainable energy production and reducing environmental degradation.

4.9 Conclusion

Agro-residue-based bioethanol production presents a promising solution for sustainable energy production, environmental conservation, and rural development. By leveraging agricultural residues, bioethanol production can reduce GHG emissions, promote carbon sequestration, and decrease dependence on fossil fuels. The success of bioethanol initiatives relies on integrated systems spanning from fields to final bioethanol production, considering climate, habitat, and management impacts. Integrated agro-residue management ensures consistent feedstock supply, while understanding the microbiome's role in soil health and plant productivity informs strategies for optimizing agro-residue production and minimizing environmental impacts. Sustainable agricultural practices, such as crop rotation and cover cropping, enhance soil conservation and fertility, reducing nitrogen oxide emissions. LCA and technoeconomic models evaluate environmental and economic performance, guiding policymakers and stakeholders. Engaging stakeholders, emphasizing environmental benefits and opportunities, promotes sustainable bioenergy production, incentivizing sustainable practices, rural development, and reduced fossil fuel dependence. Government policies and initiatives, such as the Pradhan Mantri JI-VAN Yojana and National Biofuel Policy (2018), support bioethanol production, targeting 20% ethanol blending in gasoline by 2025. Future prospects include advancements in biotechnology and process optimization, increased adoption of sustainable agricultural practices, growing demand for biofuels and renewable energy, and potential for rural development and job creation. Agro-residue-based bioethanol production has the potential to transform waste into a valuable resource, promoting energy security, environmental sustainability, and economic growth, making it a vital component of India's renewable energy landscape.

References

Abbasi, K.R., Hussain, K., Radulescu, M., Ozturk, I., 2021. Does natural resources depletion and economic growth achieve the carbon neutrality target of the UK? A way forward towards sustainable development. Resour. Policy 03014207. 74, 102341. doi:10.1016/j.resourpol.2021.102341.

Akpan, U.G., Kovo, A.S., Abdullahi, M., Ijah, J.J., 2005. The production of ethanol from maize cobs and groundnut shells. AU J. Technol. 9 (2), 106–110.

Akshaya, V., Akila, I., Murali, R., Raajasubramanian, D., Kuppan, N., Srinivasan, S., 2023. Rice Straw Biomass and Agricultural Residues as Strategic Bioenergy: effects on the Environment and Economy Path with New Directions. Energy, Environment, and Sustainability 1, 139–164. Available from: https://www.springer.com/series/15901. doi:10.1007/978-981-99-3002-9_9.

Arumugam, R., Manikandan, M., 2011. Fermentation of pretreated hydrolyzates of banana and Mango fruit wastes for ethanol production. Asian J. Exp. Biol. Sci. 2 (2), 246–256.

Arun, J., Gopinath, K.P., SundarRajan, P.S., JoselynMonica, M., Felix, V., 2019. Co-liquefaction of Prosopis juliflora with polyolefin waste for production of high grade liquid hydrocarbons. Bioresour. Technol. 18732976. 274, 296–301. Available from: www.elsevier.com/locate/biortech. doi:10.1016/j.biortech.2018.11.102.

Baig, M., Dharmadhikari, S., 2014. Bioethanol Production from Enzymatically Hydrolysed Cotton Stalk: one Approach Towards Sustainable Energy Development. Curr. World Environ. 09734929. 9 (3), 940–946. doi:10.12944/CWE.9.3.46.

Bakari, R., Asha, R., Hossein, M., Xiao Huang, N.F.I., Liew, R.K., Narayan, M., Lam, S.S., Sarma, H., 2024. Converting food waste to biofuel: a sustainable energy solution for Sub-Saharan Africa. Sustain. Chem. Environ. 29498392. 7, 100126. doi:10.1016/j.scenv.2024.100126.

Ballesteros, M., Oliva, J.M., Negro, M.J., Manzanares, P., Ballesteros, I., 2004. Ethanol from lignocellulosic materials by a simultaneous saccharification and fermentation process (SFS) with Kluyveromyces marxianus CECT 10875. Process Biochem. 13595113. 39 (12), 1843–1848. doi:10.1016/j.procbio.2003.09.011.

Beagle, E., Belmont, E., 2019. Comparative life cycle assessment of biomass utilization for electricity generation in the European Union and the United States. Energy Policy. 03014215. 128, 267–275. doi:10.1016/j.enpol.2019.01.006.

Belal, E.B., 2013. Bioethanol production from rice straw residues. Braz. J. Microbiol. 16784405. 44 (1), 225–234. Available from: https://www.scielo.br/pdf/bjm/v44n1/v44n1a33.pdf. doi:10.1590/S1517-83822013000100033Egypt.

Bhatt, S.M., Shilpa, 2014. Bioethanol production from economical agro waste (Groundnut Shell) in SSF mode. Research Journal of Pharmaceutical, Biological and Chemical Sciences 09758585. 5 (6), 1210–1218. Available from: https://www.rjpbcs.com/pdf/2014_5(6)/[175].pdf.

Bhuvaneshwari, S., Hettiarachchi, H., Meegoda, J.N., 2019. Crop Residue Burning in India: policy Challenges and Potential Solutions. Int. J. Environ. Res. Public Health. 1660-4601. 16 (5), 832. doi:10.3390/ijerph16050832.

Caffrey, K.R., Veal, M.W., 2013. Conducting an agricultural life cycle assessment: challenges and perspectives. Sci. World J. 1537744X. 2013. doi:10.1155/2013/472431.

Chandel, A.K., Kapoor, R.K., Singh, A., Kuhad, R.C., 2007. Detoxification of sugarcane bagasse hydrolysate improves ethanol production by Candida shehatae NCIM 3501. Bioresour. Technol. 09608524. 98 (10), 1947–1950. doi:10.1016/j.biortech.2006.07.047.

Chen, S., Feng, H., Zheng, J., Jianguo, Y., Song, Y., Yang, H., Zhou, M., 2020. Life Cycle Assessment and Economic Analysis of Biomass Energy Technology in China: a Brief Review. Processes. 2227-9717. 8 (9), 1112. doi:10.3390/pr8091112.

Cheng, K.K., Cai, B.Y., Zhang, J.A., Ling, H.Z., Zhou, Y.J., Ge, J.P., Xu, J.M., 2008. Sugarcane bagasse hemicellulose hydrolysate for ethanol production by acid recovery process. Biochem. Eng. J. 1369703X. 38 (1), 105–109. doi:10.1016/j.bej.2007.07.012.

Damayanti, D., Supriyadi, D., Amelia, D., Saputri, D.R., Devi, Y.L.L., Auriyani, W.A., Wu, H.S., 2021. Conversion of lignocellulose for bioethanol production, applied in bio–polyethylene terephthalate. Polymers. 20734360. 13 (17). Available from: https://www.mdpi.com/2073-4360/13/17/2886/pdf. doi:10.3390/polym13172886.

Ding, T.Y., Hii, S.L., Ong, L.G.A., 2012. Comparison of pretreatment strategies for conversion of coconut husk fiber to fermentable sugars. BioResources. 19302126. 7 (2), 1540–1547. Available from: https://www.ncsu.edu/bioresources/BioRes_07/BioRes_07_2_1540_correction_Ong_DH_Compar_Pretreat_Strat_Coconut_Husk_Sugars_2491.pdf. doi:10.15376/biores.7.2.1540-1547.

Duque, A., Álvarez, C., Doménech, P., Manzanares, P., Moreno, A.D., 2021. Advanced Bioethanol Production: from Novel Raw Materials to Integrated Biorefineries. Processes. 2227-9717. 9 (2), 206. doi:10.3390/pr9020206.

Ebrahimi, M., Caparanga, A.R., Ordono, E.E., Villaflores, O.B., 2017. Evaluation of organosolv pretreatment on the enzymatic digestibility of coconut coir fibers and bioethanol production via simultaneous saccharification and fermentation. Renew. Energy. 18790682. 109, 41–48. Available from: https://www.journals.elsevier.com/renewable-and-sustainable-energy-reviews/. doi:10.1016/j.renene.2017.03.011.

El-Araby, R., 2024. Biofuel production: exploring renewable energy solutions for a greener future. Biotechnol. Biofuels Bioprod. 2731-3654. 17 (1), doi:10.1186/s13068-024-02571-9.

Enebe, N.L., Chigor, C.B., Obileke, K., Lawal, M.S., Enebe, M.C., 2023. Biogas and Syngas Production from Sewage Sludge: a Sustainable Source of Energy Generation. Methane. 2674-0389. 2 (2), 192–217. doi:10.3390/methane2020014.

Food and Agriculture Organization of the United Nations. (n.d.). Agricultural production statistics 2010-2023. Retrieved from: https://www.fao.org/statistics/highlights-archive/highlights-detail/agricultural-production-statistics-2010-2023/en.

Gatdula, K.M., Blaquera, M.S.F.N., Jimena, C.M.V., Elegado, F.B., Alcantara, J.Z., Guerrero, G.A.M., 2021. A comparative study on bioethanol production from rice straw and banana pseudostem through simultaneous saccharification and fermentation using Kluyveromyces marxianus. IOP Conf. Ser. Earth Environ. Sci. 1755-1307. 765 (1), 012006. doi:10.1088/1755-1315/765/1/012006.

Gautam, S., Bora, B., Dutta, D., Tripathi, A.D., Srivastava, J., Thatoi, H.N., Srivastava, S.K., Khade, S.M., Geed, S.R., 2026. Integrated biorefinery approaches for the sustainable valorization of agricultural residues into biofuels, bioplastics, and bioactive compounds. Sustain. Chem. Clim. Action 27728269. 8, 100173. doi:10.1016/j.scca.2025.100173.

Gonçalves, F.A., Ruiz, H.A., Dos Santos, E.S., Teixeira, J.A., de Macedo, G.R., 2016. Bioethanol production by Saccharomyces cerevisiae, Pichia stipitis, and Zymomonas mobilis from delignified mature coconut fibre with lignin extraction according to the biorefinery concept. Renew. Energy 94, 353–365. doi:https://doi.org/10.1016/j.renene.2016.03.045.

Gonçalves, F.A., Ruiz, H.A., Nogueira, C.D.C., Santos, E.S.D., Teixeira, J.A., De Macedo, G.R., 2014. Comparison of delignified coconuts waste and cactus for fuel-ethanol production by the simultaneous and semi-simultaneous saccharification and fermentation strategies. Fuel. 00162361. 131, 66–76. doi:10.1016/j.fuel.2014.04.021.

Gonçalves, F.A., Ruiz, H.A., Santos, E.S.D., Teixeira, J.A., De Macedo, G.R., 2015. Bioethanol production from coconuts and cactus pretreated by autohydrolysis. Ind. Crops Prod. 09266690. 77, 1–12. Available from: www.elsevier.com/inca/publications/store/5/2/2/8/2/5. doi:10.1016/j.indcrop.2015.06.041.

Guerrero, A.B., Ballesteros, I., Ballesteros, M., 2018. The potential of agricultural banana waste for bioethanol production. Fuel. 00162361. 213, 176–185. Available from: https://www.journals.elsevier.com/fuel/. doi:10.1016/j.fuel.2017.10.105.

Han, W., Liu, Y., Xiaobin, X., Huang, J., Hong, H., Chen, L., Qiu, S., Tang, J., Hou, P., 2020. Bioethanol production from waste hamburger by enzymatic hydrolysis and fermentation. J. Clean. Prod. 09596526. 264, 121658. doi:10.1016/j.jclepro.2020.121658.

Harish, K.R.Y., Srijana, M., Madhusudhan, R.D., Gopal, R., 2010. Coculture fermentation of banana agro-waste to ethanol by cellulolytic thermophilic Clostridium thermocellum CT2. Afr. J. Biotechnol. 1684-5315. 9 (13), 1926–1934. doi:10.5897/AJB09.1217.

Hasan, A.H., Avami, A., 2021. Comparative assessment of bioethanol supply chain: insights from Iran. Biofuels. 1759-7269. 12 (4), 475–483. doi:10.1080/17597269.2018.1496385.

Hernández-Salas, J.M., Villa-Ramírez, M.S., Veloz-Rendón, J.S., Rivera-Hernández, K.N., González-César, R.A., Plascencia-Espinosa, M.A., Trejo-Estrada, S.R., 2009. Comparative hydrolysis and fermentation of sugarcane and agave bagasse. Bioresour. Technol. 09608524. 100 (3), 1238–1245. doi:10.1016/j.biortech.2006.09.062.

Hernawan, H., Maryana, R., Pratiwi, D., Wahono, S.K., Darsih, C., Hayati, S.N., Poeloengasih, C.D., Nisa, K., Indrianingsih, A.W., Prasetyo, D.J., Jatmiko, T.H., Kismurtono, M., Rosyida, V.T., 2017. Bioethanol production from sugarcane bagasse by simultaneous sacarification and fermentation using Saccharomyces cerevisiae. AIP Conf. Proc. 15517616. 1823. Available from: https://scitation.aip.org/content/aip/proceeding/aipcp. doi:10.1063/1.4978099.

Jiang, W., Chang, S., Hongqiang, L., Oleskowicz-Popiel, P., Jian, X., 2015. Liquid hot water pretreatment on different parts of cotton stalk to facilitate ethanol production. Bioresour. Technol. 09608524. 176, 175–180. doi:10.1016/j.biortech.2014.11.023.

Joseph, A.M., Tulasi, Y., Shrivastava, D., Kiran, B., 2023. Techno-economic feasibility and exergy analysis of bioethanol production from waste. Energy Convers. Manag. X. 25901745. 18, 100358. doi:10.1016/j.ecmx.2023.100358.

Kaur, U., Oberoi, H.S., Bhargav, V.K., Sharma-Shivappa, R., Dhaliwal, S.S., 2012. Ethanol production from alkali- and ozone-treated cotton stalks using thermotolerant Pichia kudriavzevii HOP-1. Ind. Crops Prod. 09266690. 37 (1), 219–226. doi:10.1016/j.indcrop.2011.12.007.

Kazi, F.K., Fortman, J.A., Anex, R.P., Hsu, D.D., Aden, A., Dutta, A., Kothandaraman, G., 2010. Techno-economic comparison of process technologies for biochemical ethanol production from corn stover. Fuel. 00162361. 89 (1), S20. doi:10.1016/j.fuel.2010.01.001.

Kazmi, A., Sultana, T., Ali, A., Nijabat, A., Gaojie, L., Hou, H., 2025. Innovations in bioethanol production: a comprehensive review of feedstock generations and technology advances. Energy Strateg. Rev. 2211467X. 57, 101634. doi:10.1016/j.esr.2024.101634.

Kedong, M., Ruan, Z., 2015. Production of a lignocellulolytic enzyme system for simultaneous bio-delignification and saccharification of corn stover employing co-culture of fungi. Bioresour. Technol. 09608524. 175, 586–593. doi:10.1016/j.biortech.2014.10.161.

Khounani, Z., Nazemi, F., Shafiei, M., Aghbashlo, M., Tabatabaei, M., 2019. Techno-economic aspects of a safflower-based biorefinery plant co-producing bioethanol and biodiesel. Energy Convers. Manag. 01968904. 201, 112184. doi:10.1016/j.enconman.2019.112184.

Konti, A., Damigos, D., 2018. Exploring strengths and weaknesses of bioethanol production from bio-waste in Greece using Fuzzy Cognitive Maps. Energy Policy. 03014215. 112, 4–11. doi:10.1016/j.enpol.2017.09.053.

Kumar, A., Singh, J., Baskar, C., 2019. Lignocellulosic biomass for bioethanol production through microbes: strategies to improve process efficiency. In: Pandey, A., Larroche, C., Ricke, S.C., Dussap, C.-G., Gnansounou, E. (Eds.) Prospects of Renewable Bioprocessing in Future Energy Systems. Elsevier, pp. 357–386.

Kumar, B., Verma, P., 2020. Application of hydrolytic enzymes in biorefinery and its future prospects. In: Srivastava, N., Srivastava, M., Mishra, P.K., Gupta, V.K., Eds. Microbial Strategies for Techno-economic Biofuel Production. Springer, pp. 59–86. doi:https://doi.org/10.1007/978-981-15-7190-9_3.

Liu, Z.H., Chen, H.Z., 2016. Simultaneous saccharification and co-fermentation for improving the xylose utilization of steam exploded corn stover at high solid loading.

Bioresour. Technol. 18732976. 201, 15–26. Available from: www.elsevier.com/locate/biortech. doi:10.1016/j.biortech.2015.11.023.

Luo, L., van der Voet, E., Huppes, G., 2010. Energy and Environmental Performance of Bioethanol from Different Lignocelluloses. Int. J. Chem. Eng. 1687-806X. 2010, 1–12. doi:10.1155/2010/740962.

Malik, K., Capareda, S.C., Kamboj, B.R., Malik, S., Singh, K., Arya, S., Bishnoi, D.K., 2024. Biofuels Production: a Review on Sustainable Alternatives to Traditional Fuels and Energy Sources. Fuels. 2673-3994. 5 (2), 157–175. doi:10.3390/fuels5020010.

Mankar, A.R., Pandey, A., Arindam Modak, K.K.P., 2021. Pretreatment of lignocellulosic biomass: a review on recent advances. Bioresour. Technol. 09608524. 334, 125235. doi:10.1016/j.biortech.2021.125235.

MARTIN, C., MARCET, M., ALMAZAN, O., JONSSON, L., 2007. Adaptation of a recombinant xylose-utilizing Saccharomyces cerevisiae strain to a sugarcane bagasse hydrolysate with high content of fermentation inhibitors. Bioresour. Technol. 09608524. 98 (9), 1767–1773. doi:10.1016/j.biortech.2006.07.021.

Matharasi, A., Uma, C., Sivagurunathan, P., Sampathkumar, P., 2018. Determination of bio-ethanol potential from banana waste using indigenous yeast (Saccharomyces cerevisiae KX033583). J. Pharmacogn. Phytochem. 7 (5), 1406–1410.

Meena, R.K., Sahu, R., Shukla, P., Thakur, S., 2015. Bio-ethanol production from ligno-cellulosic banana waste using co-culture techniques. Curr. Trends Biotechnol. Pharm. 22307303. 9 (3), 259–265. Available from: www.abap.co.in.

Ministry of Agriculture and Farmers Welfare. (n.d.). State-wise agricultural production data. Retrieved from: https://upag.gov.in/dashreports/statewiseapy?rtab=Area%2C+Production+%26+Yield&rtype=reports.

Ministry of Petroleum and Natural Gas (MoPNG), Government of India. (2018). National Policy on Biofuels. Government of India. Retrieved July 12, 2025, from https://biofuel.rajasthan.gov.in/pdf/National%20biofuel_policy.pdf.

Miret, C., Chazara, P., Montastruc, L., Negny, S., Domenech, S., 2016. Design of bioethanol green supply chain: comparison between first and second generation biomass concerning economic, environmental and social criteria. Comput. Chem. Eng. 00981354. 85, 16–35. doi:10.1016/j.compchemeng.2015.10.008.

Mirmohamadsadeghi, S., Chen, Z., Wan, C., 2016. Reducing biomass recalcitrance via mild sodium carbonate pretreatment. Bioresour. Technol. 09608524. 209, 386–390. doi:10.1016/j.biortech.2016.02.096.

Nwaze, I.O., Emmanuel, A., Akubueze, E.U., 2023. Suitability of groundnut shells for bio-ethanol production using Saccharomyces cerevisiae. J. Mater. Sci. Res. Rev. 11 (2), 1–9.

Nyachaka, C.J., Yawas, D.S., Pam, G.Y., 2013. Production and performance evaluation of bioethanol fuel from groundnut shell waste. Am. J. Eng. Res. 2 (10), 225–232.

Oberoi, H.S., Vadlani, P.V., Saida, L., Bansal, S., Hughes, J.D., 2011. Ethanol production from banana peels using statistically optimized simultaneous saccharification and fermentation process. Waste Manag. 0956053X. 31 (7), 1576–1584. doi:10.1016/j.wasman.2011.02.007.

Öhgren, K., Bengtsson, O., Gorwa-Grauslund, M.F., Galbe, M., Hahn-Hägerdal, B., Zacchi, G., 2006. Simultaneous saccharification and co-fermentation of glucose and xylose in steam-pretreated corn stover at high fiber content with Saccharomyces cerevisiae TMB3400. J. Biotechnol. 01681656. 126 (4), 488–498. doi:10.1016/j.jbiotec.2006.05.001.

Öhgren, K., Bura, R., Lesnicki, G., Saddler, J., Zacchi, G., 2007. A comparison between simultaneous saccharification and fermentation and separate hydrolysis and fermentation using steam-pretreated corn stover. Process Biochem. 13595113. 42 (5), 834–839. doi:10.1016/j.procbio.2007.02.003.

Oyegoke, T., Sardauna, M.Y., Obadiah, E., Abubakar, H.A., 2021. Exploration of biomass for the production of bioethanol: a process modelling and simulation study. Renew. Energy Res. Appl. J. 2 (1), 51–69. doi:https://doi.org/10.22044/RERA.2020.10287.1042.

Palacios, A.S., Ilyina, A., Ramos-González, R., Aguilar, C.N., Martínez-Hernández, J.L., Segura-Ceniceros, E.P., González, M.L.C., Aguilar, M., Ballesteros, M., Oliva, J.M., Ruiz, H.A., 2021. Ethanol production from banana peels at high pretreated substrate loading: comparison of two operational strategies. Biomass Convers. Biorefin. 21906823. 11 (5), 1587–1596. Available from: https://www.springer.com/engineering/energy+technology/journal/13399. doi:10.1007/s13399-019-00562-7.

Palacios, S., Ruiz, H.A., Ramos-Gonzalez, R., Martínez, J., Segura, E., Aguilar, M., Aguilera, A., Michelena, G., Aguilar, C., Ilyina, A., 2017. Comparison of physicochemical pretreatments of banana peels for bioethanol production. Food Sci. Biotechnol. 12267708. 26 (4), 993–1001. Available from: https://www.springerlink.com/content/n3031612t262/. doi:10.1007/s10068-017-0128-9.

Panahi, H.K.S., Dehhaghi, M., Guillemin, G.J., Gupta, V.K., Lam, S.S., Aghbashlo, M., Tabatabaei, M., 2022. Bioethanol production from food wastes rich in carbohydrates. Curr. Opin. Food Sci. 22147993. 43, 71–81. Available from: https://www.journals.elsevier.com/current-opinion-in-food-science/. doi:10.1016/j.cofs.2021.11.001.

Phwan, C.K., Chew, K.W., Sebayang, A.H., Ong, H.C., Ling, T.C., Malek, M.A., Ho, Y.C., Show, P.L., 2019. Effects of acids pre-treatment on the microbial fermentation process for bioethanol production from microalgae. Biotechnol. Biofuels. 17546834. 12 (1). Available from: https://www.biotechnologyforbiofuels.com/. doi:10.1186/s13068-019-1533-5.

Polipalli, K., Suraboyina, S., Kashimalla, M., Polumati, A., 2025. A review on value addition of Agricultural Residues by Chemical and Bio-chemical Processes to abate environmental pollution. Green Technol. Sustain. 29497361. 3 (4), 100241. doi:10.1016/j.grets.2025.100241.

Press Information Bureau (PIB), Government of India; Ministry of Petroleum and Natural Gas (MoPNG). (2025). Ethanol Blended Petrol (EBP) Programme in India. Government of India. Retrieved October 1, 2025, from https://www.pib.gov.in/PressReleasePage.aspx?PRID=2097307®=3&lang=2.

Press Information Bureau (PIB), Government of India. (2022). Press Release. Press Information Bureau. Retrieved August 17, 2025, from https://www.pib.gov.in/PressReleasePage.aspx?PRID=1843441®=3&lang=2.

Qing, Q., Zhou, L., Guo, Q., Gao, X., Zhang, Y., He, Y., Zhang, Y., 2017. Mild alkaline presoaking and organosolv pretreatment of corn stover and their impacts on corn stover composition, structure, and digestibility. Bioresour. Technol. 18732976. 233, 284–290. Available from: www.elsevier.com/locate/biortech. doi:10.1016/j.biortech.2017.02.106.

Quintero, J.A., Moncada, J., Cardona, C.A., 2013. Techno-economic analysis of bioethanol production from lignocellulosic residues in Colombia: a process simulation approach. Bioresour. Technol. 18732976. 139, 300–307. Available from: www.elsevier.com/locate/biortech. doi:10.1016/j.biortech.2013.04.048.

Rabah, A.B., Oyeleke, S.B., Manga, S.B., Hassan, L.G., 2011. Microbial pretreatment of rice husk and groundnut shell for bioethanol production. Int. Res. J. Microbiol. 2 (8), 255–260.

Raman, R., Sreenivasan, A., Kulkarni, N.V., Suresh, M., Nedungadi, P., 2025. Analyzing the contributions of biofuels, biomass, and bioenergy to sustainable development goals. iScience. 25890042. 28 (4), 112157. doi:10.1016/j.isci.2025.112157.

Ramanathan, A., Begum, K.M.M.S., Pereira, A.O., Cohen, C., 2022. Pyrolysis of waste biomass: toward sustainable development. A Thermo-economic Approach to Energy from Waste. doi:10.1016/B978-0-12-824357-2.00005-X.

Rațu, R.N., Veleșcu, I.D., Stoica, F., Usturoi, A., Arsenoaia, V.N., Crivei, I.C., Postolache, A.N., Lipşa, F.D., Filipov, F., Florea, A.M., Chițea, M.A., Brumă, I.S., 2023. Application of Agri-Food By-Products in the Food Industry. Agriculture (Switzerland.

20770472. 13 (8). Available from: www.mdpi.com/journal/agriculture. doi:10.3390/agriculture13081559.

Rezania, S., Oryani, B., Cho, J., Talaiekhozani, A., Sabbagh, F., Hashemi, B., Rupani, P.F., Mohammadi, A.A., 2020. Different pretreatment technologies of lignocellulosic biomass for bioethanol production: an overview. Energy. 03605442. 199. Available from: https://www.journals.elsevier.com/energy. doi:10.1016/j.energy.2020.117457.

Salaheldeen, M., Mariod, A.A., Aroua, M.K., Rahman, S.M.A., Soudagar, M.E.M., Fattah, I.M.R., 2021. Current state and perspectives on transesterification of triglycerides for biodiesel production. Catalysts. 20734344. 11 (9). Available from: https://www.mdpi.com/2073-4344/11/9/1121/pdf. doi:10.3390/catal11091121.

Saleem, M., 2022. Possibility of utilizing agriculture biomass as a renewable and sustainable future energy source. Heliyon. 24058440. 8 (2), e08905. doi:10.1016/j.heliyon.2022.e08905.

Serrano-Echeverry, V.A., Guerrero-Fajardo, C.A., Castro-Tibabisco, K.T., 2025. A Review of Biobutanol: eco-Friendly Fuel of the Future—History, Current Advances, and Trends. Fuels. 2673-3994. 6 (3), 55. doi:10.3390/fuels6030055.

Shankar, K., Kulkarni, N.S., Rakesh Sajjanshetty, S.K.J., Sreeramulu, K., 2020. Co-production of xylitol and ethanol by the fermentation of the lignocellulosic hydrolysates of banana and water hyacinth leaves by individual yeast strains. Ind. Crops Prod. 09266690. 155, 112809. doi:10.1016/j.indcrop.2020.112809.

Singh, A., Bajar, S., Bishnoi, N.R., 2017. Physico-chemical pretreatment and enzymatic hydrolysis of cotton stalk for ethanol production by Saccharomyces cerevisiae. Bioresour. Technol. 18732976. 244, 71–77. Available from: www.elsevier.com/locate/biortech. doi:10.1016/j.biortech.2017.07.123.

Singh, D.P., Prabha, R., 2018. Bioconversion of Agricultural Wastes into High Value Biocompost: a Route to Livelihood Generation for Farmers. Adv. Recycl. Waste Manag. 24757675. 02 (03), doi:10.4172/2475-7675.1000137.

Singhania, R.R., Saini, J.K., Saini, R., Adsul, M., Mathur, A., Gupta, R., Tuli, D.K., 2014. Bioethanol production from wheat straw via enzymatic route employing penicillium janthinellum cellulases. Bioresour. Technol. 18732976. 169, 490–495. Available from: www.elsevier.com/locate/biortech. doi:10.1016/j.biortech.2014.07.011.

Smith, L.G., Kirk, G.J.D., Jones, P.J., Williams, A.G., 2019. The greenhouse gas impacts of converting food production in England and Wales to organic methods. Nat. Commun. 2041-1723. 10 (1), doi:10.1038/s41467-019-12622-7.

Soares, J., Demeke, M.M., Van de Velde, M., Foulquié-Moreno, M.R., Kerstens, D., Sels, B.F., Verplaetse, A., Fernandes, A.A.R., Thevelein, J.M., Fernandes, P.M.B., 2017. Fed-batch production of green coconut hydrolysates for high-gravity second-generation bioethanol fermentation with cellulosic yeast. Bioresour. Technol. 18732976. 244, 234–242. Available from: www.elsevier.com/locate/biortech. doi:10.1016/j.biortech.2017.07.140.

Soares, J., Ruiz, H.A., Silva, D.P., Teixeira, J.A., Gonçalves, F.A., 2016. Green coconut mesocarp pretreated by an alkaline process as raw material for bioethanol production. Bioresour. Technol. 216, 744–753. doi:https://doi.org/10.1016/j.biortech.2016.05.105.

Song, L., Fuying, M., Zeng, Y., Zhang, X., Hongbo, Y., 2013. The promoting effects of manganese on biological pretreatment with Irpex lacteus and enzymatic hydrolysis of corn stover. Bioresour. Technol. 09608524. 135, 89–92. doi:10.1016/j.biortech.2012.09.004.

Suhag, M., Kumar, A., Singh, J., 2020. Saccharification and fermentation of pretreated banana leaf waste for ethanol production. SN Appl. Sci. 2523-3963. 2 (8), doi:10.1007/s42452-020-03215-x.

Talebnia, F., Karakashev, D., Angelidaki, I., 2010. Production of bioethanol from wheat straw: an overview on pretreatment, hydrolysis and fermentation. Bioresour. Technol. 09608524. 101 (13), 4744–4753. doi:10.1016/j.biortech.2009.11.080.

Tanquilut, M.R., Elauria, J., Amongo, R.M., Suministrado, D., Yaptenco, K., Elauria, M., 2019. Biomass Characterization of Pigeon Pea (Cajanus cajan) Wood for Thermochemical Conversion. Philipp. J. Agric. Biosyst. Eng. 0119-7312. 15 (1), 39. doi:10.48196/015.01.2019.04.

Teymouri, F., Laureano-Perez, L., Alizadeh, H., Dale, B.E., 2005. Optimization of the ammonia fiber explosion (AFEX) treatment parameters for enzymatic hydrolysis of corn stover. Bioresour. Technol. 09608524. 96 (18), 2014–2018. Available from: www.elsevier.com/locate/biortech. doi:10.1016/j.biortech.2005.01.016.

Thatoi, H., Dash, P.K., Mohapatra, S., Swain, M.R., 2016. Bioethanol production from tuber crops using fermentation technology: a review. Int. J. Sustain. Energy. 1478646X. 35 (5), 443–468. Available from: http://www.tandfonline.com/toc/gsol20/current. doi:10.1080/14786451.2014.918616.

Thuppahige, R.T.W., Barner, L., Shahbazi, M., Fraga, G., Moghaddam, L., 2025. A comprehensive review of sustainable valorisation of lignocellulosic biomass and plastic waste into biofuels and chemicals via co-liquefaction. Waste Manag. 0956053X. 202, 114827. doi:10.1016/j.wasman.2025.114827.

Uppugundla, N., Sousa, L.D.C., Chundawat, S.P.S., Yu, X., Simmons, B., Singh, S., Gao, X., Kumar, R., Wyman, C.E., Dale, B.E., Balan, V., 2014. A comparative study of ethanol production using dilute acid, ionic liquid and AFEX™ pretreated corn stover. Biotechnol. Biofuels 17546834. 7 (1), doi:10.1186/1754-6834-7-72.

Vaithanomsat, P., Apiwatanapiwat, W., Chumchuent, N., Kongtud, W., Sundhrarajun, S., 2011. The potential of coconut husk utilization for bioethanol production. Kasetsart J. Nat. Sci. 00755192. 45 (1), 159–164.

Velvizhi, G., Jacqueline, P.J., Shetti, N.P., Latha, K., Mohanakrishna, G., Aminabhavi, T.M., 2023. Emerging trends and advances in valorization of lignocellulosic biomass to biofuels. J. Environ. Manag. 03014797. 345, 118527. doi:10.1016/j.jenvman.2023.118527.

Wang, J., Azam, W., 2024. Natural resource scarcity, fossil fuel energy consumption, and total greenhouse gas emissions in top emitting countries. Geosci. Front. 16749871. 15 (2), 101757. doi:10.1016/j.gsf.2023.101757.

Wang, M., Zhou, D., Wang, Y., Wei, S., Yang, W., Kuang, M., Ma, L., Fang, D., Xu, S., Du, S.K., 2016. Bioethanol production from cotton stalk: a comparative study of various pretreatments. Fuel. 00162361. 184, 527–532. Available from: https://www.journals.elsevier.com/fuel/. doi:10.1016/j.fuel.2016.07.061.

Wi, S.G., Choi, I.S., Kim, K.H., Kim, H.M., Bae, H.J., 2013. Bioethanol production from rice straw by popping pretreatment. Biotechnol. Biofuels 17546834. 6 (1), doi:10.1186/1754-6834-6-166.

Yoswathana, 2010. Bioethanol Production from Rice Straw. Energy Res. J. 1949-0151. 1 (1), 26–31. doi:10.3844/erjsp.2010.26.31.

Yusuf, A.A., Inambao, F.L., 2019. Bioethanol production techniques from lignocellulosic biomass as alternative fuel: a review. Int. J. Adv. Res. Eng. Technol. 09766499. 10 (3), 259–288. Available from: http://www.iaeme.com/MasterAdmin/Journal_uploads/IJARET/VOLUME_10_ISSUE_3/IJARET_10_03_026.pdf. doi:10.34218/IJARET.10.3.2019.026.

Zhao, L., Zhang, X., Jie, X., Xunmin, O., Chang, S., Maorong, W., 2015. Techno-Economic Analysis of Bioethanol Production from Lignocellulosic Biomass in China: dilute-Acid Pretreatment and Enzymatic Hydrolysis of Corn Stover. Energies. 1996-1073. 8 (5), 4096–4117. doi:10.3390/en8054096.

CHAPTER 5

Biodiesel production from agricultural residues

Kotteeswaran Santhanam[a], A. Mohan[b], and D. Gayathri[c]
[a]Department of Civil Engineering, SRM Institute of Science and Technology, Ramapuram, Chennai, Tamil Nadu, India
[b]Department of Civil Engineering, Easwari Engineering College, Ramapuram, Chennai, Tamil Nadu, India
[c]Department of Computer Science and Engineering, Velammal Engineering College, Surapet, Chennai, India

5.1 Overview of biodiesel production process

Biodiesel is a renewable, biodegradable fuel derived from organic lipids such as vegetable oils or animal fats. When agricultural residues such as nonedible oil seeds, waste cooking oil, or lignocellulosic biomass are used, it offers an eco-friendly alternative that doesn't compete with food crops. The production process typically involves extraction of oil, followed by transesterification: a chemical process that converts triglycerides into methyl esters (biodiesel) and glycerol using an alcohol (usually methanol) and a catalyst. Biodiesel can be used in standard diesel engines without significant modification and reduces emissions of particulate matter, hydrocarbons, carbon monoxide, and sulfur oxides. Its sustainability is enhanced when residues and wastes are used as feedstock (Fig. 5.1).

5.2 Feedstock selection and preparation

5.2.1 Types of agricultural residues

Agricultural residues suitable for biodiesel production come in several forms, each with distinct characteristics and processing requirements. Nonedible oilseeds such as *Jatropha*, *Karanja*, and *Mahua* are widely used, especially in rural and semi-arid regions, due to their high oil content and the advantage of not competing with food crops. These oilseeds can be cultivated on marginal lands and are often part of sustainable land-use strategies. Waste cooking oils represent another valuable category, primarily collected from urban areas with dense populations and a vibrant food service industry (Babin et al., 2021). They offer a low-cost, readily available feedstock, though pretreatment is often required to remove impurities and reduce high free fatty acid (FFA) levels.

Bioenergy from Agricultural Residues
https://doi.org/10.1016/B978-0-443-24726-2.00006-7

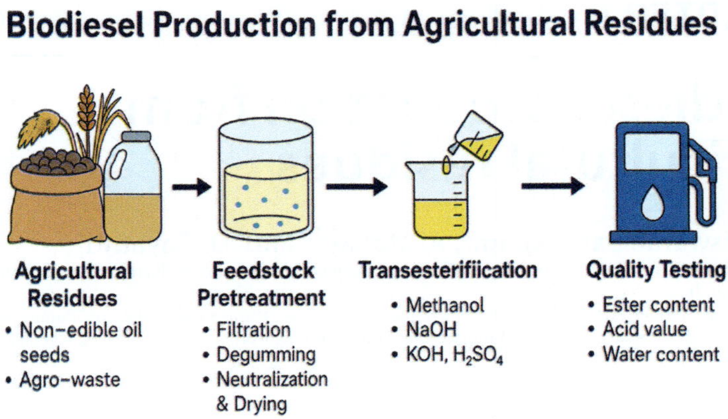

Fig. 5.1 Process of biodiesel production from agricultural residues.

Agro-industrial residues, including *rice bran oil*, *cottonseed oil*, and *corn oil* recovered from distiller's grains in ethanol plants, are by-products of agricultural processing industries. These oils are often underutilized and can be diverted to biodiesel production with relatively minimal modification. Lastly, lignocellulosic biomass, such as crop stalks, husks, and bagasse, represents a more complex feedstock. Although it is abundant, it requires thermochemical conversion processes such as pyrolysis or gasification to produce biodiesel intermediates such as bio-oil or syngas, which are then upgraded to fuel. The diversity of these residues provides flexibility in feedstock selection, enabling localized and cost-effective biodiesel production strategies tailored to regional agricultural profiles.

5.2.2 Feedstock criteria

The selection of feedstock for biodiesel production is guided by several important criteria that directly influence the efficiency, cost-effectiveness, and scalability of the process. Oil content is one of the most critical parameters; feedstocks with higher oil yield per unit mass are preferred because they result in greater biodiesel output and reduce the overall processing cost (Bachanek et al., 2021). For example, nonedible oilseeds such as Jatropha and Mahua offer substantial oil yields, making them attractive for commercial production.

FFA levels also play a vital role in determining the suitability of a feedstock. When FFA levels exceed 2%, direct base-catalyzed transesterification becomes problematic, leading to soap formation that reduces biodiesel yield and complicates separation. In such cases, an acid-catalyzed pretreatment step is essential to esterify FFAs before the main reaction. This adds to processing complexity and cost, making FFA content a key factor in economic feasibility.

Availability and seasonality of the feedstock affect the consistency of production. Agricultural residues are often seasonal, tied to harvest cycles, and may not be available year-round. Ensuring a steady supply requires strategic planning, storage solutions, or blending of multiple feedstocks to maintain operations without interruption (Bakkaloglu and Hawkes, 2024). Lastly, cost and logistics significantly influence the overall economics of biodiesel production. Feedstocks located far from processing facilities incur higher transportation costs, especially when dealing with bulky or perishable residues. Therefore, local availability and proximity to processing infrastructure are advantageous, supporting decentralized production models and improving overall project viability.

5.2.3 Pretreatment of feedstock

Before undergoing transesterification, agricultural residue-based oils must be properly pretreated to ensure a smooth and efficient conversion process. One of the initial steps is filtration, which is used to eliminate solid particles such as dust, husk fragments, or food residues (in the case of used cooking oils). This step also helps remove excess moisture, which can interfere with the chemical reaction and promote soap formation. Following filtration, the oil often undergoes degumming, a process that removes phospholipids, also known as gums. These compounds can severely reduce the effectiveness of catalysts and may cause foaming or emulsification during biodiesel processing. Degumming is especially important for oils derived from agro-industrial residues such as rice bran or cottonseed oil.

If the feedstock has a high FFA content, it must be treated to lower its acidity. This is achieved through neutralization, where acid esterification is typically employed. In this process, FFAs react with alcohol (usually methanol) in the presence of an acid catalyst to form esters, thus preventing soap formation in the main transesterification step (Balcioglu et al., 2022). The final pretreatment step is drying, which removes residual water from the oil. Even small amounts of moisture can lead to hydrolysis, resulting in more FFAs and, consequently, more soap during the reaction. Drying is therefore essential to maintain reaction efficiency and to ensure that the biodiesel produced meets quality standards. Collectively, these pretreatment steps are vital in converting variable-quality agricultural residues into a clean and reactive oil phase suitable for biodiesel synthesis.

5.3 Transesterification process and catalysts

Transesterification is the core process in biodiesel production, involving the reaction of triglycerides (from oils) with an alcohol (methanol or ethanol) in the presence of a catalyst to form biodiesel and glycerol.

5.3.1 Chemical reaction

The chemical equation succinctly illustrates the transesterification process used in biodiesel production:

$$\text{Triglyceride} + \text{Methanol} \rightarrow \text{Biodiesel}\,(\text{Methyl Ester}) + \text{Glycerol}$$

In this reaction, triglycerides, which are the principal components of vegetable oils or animal fats, react with methanol in the presence of a catalyst. During transesterification, each triglyceride molecule splits into three FAMEs (biodiesel) and one molecule of glycerol. This conversion is crucial because the resulting biodiesel has significantly lower viscosity than the original oil, making it suitable for use in diesel engines. In addition, the production of glycerol as a byproduct adds economic value, as it can be utilized in various industrial applications ranging from cosmetics to pharmaceuticals.

5.3.2 Types of transesterification

Transesterification, the core chemical reaction in biodiesel production, can be carried out using different types of catalysts, each with its own advantages and limitations. The base-catalyzed transesterification method is the most commonly used in commercial biodiesel plants due to its high reaction rate and cost-effectiveness. It typically employs alkaline catalysts such as sodium hydroxide (NaOH) or potassium hydroxide (KOH). This method is highly efficient when the feedstock has a low FFA content, usually below 2%. However, if FFA levels are high, it leads to soap formation, which reduces biodiesel yield and complicates the separation of glycerol.

In contrast, acid-catalyzed transesterification is more suitable for feedstocks with high FFA content, such as waste cooking oils or certain nonedible oil sources. This method uses strong acids such as sulfuric acid or hydrochloric acid as catalysts. Although the reaction rate is slower compared to the base-catalyzed process, acid catalysis avoids soap formation and enables simultaneous esterification of FFAs and transesterification of triglycerides, making it ideal for low-quality or untreated oils.

A more environmentally friendly yet less commonly used approach is enzyme-catalyzed transesterification, which employs lipase enzymes. These biological catalysts operate under milder conditions—lower temperatures and neutral pH—and produce minimal waste, making the process more sustainable. In addition, enzymes can be reused multiple times, improving their long-term cost efficiency. However, the initial cost of enzymes and their slower reaction rates make this method less economically viable for large-scale operations at present, though ongoing research and advancements are making it increasingly attractive for niche and decentralized biodiesel production systems.

5.3.3 Process parameters

The efficiency and yield of the transesterification process in biodiesel production depend heavily on the careful optimization of several key operating parameters. One of the most critical factors is the molar ratio of methanol to oil, which is typically maintained at 6:1 (Barr et al., 2021). This excess of alcohol drives the reaction forward, ensuring complete conversion of triglycerides into methyl esters (biodiesel). While a higher ratio can improve yield, it may also increase costs and complicate the recovery of unused methanol.

Temperature plays a significant role in determining the reaction rate. The process is usually carried out at around 60°C, which is just below methanol's boiling point. This temperature is sufficient to speed up the reaction without causing excessive evaporation of methanol or degrading the oil. The reaction time varies depending on the catalyst and feedstock used but generally falls within 1–2 h. Base-catalyzed reactions are faster and often complete within an hour, while acid- or enzyme-catalyzed processes require more time due to slower reaction kinetics.

Another important parameter is catalyst concentration, which is typically maintained between 0.5% and 1% by weight of the oil. Using too little catalyst can result in incomplete conversion, while excessive amounts may cause emulsification or soap formation, complicating the separation of biodiesel and glycerol. Balancing these parameters is essential for achieving high conversion efficiency, minimizing side reactions, and ensuring the economic viability of biodiesel production.

5.3.4 Equipment used

Biodiesel production infrastructure involves a combination of reactor systems and downstream processing units designed to ensure efficient conversion and product purification. Two primary types of reactors are employed depending on the scale of production: batch reactors and continuous-flow reactors. Batch reactors are commonly used in small- to medium-scale biodiesel facilities due to their operational flexibility and lower initial investment. In this setup, the feedstock oil, alcohol, and catalyst are loaded into a single tank, where the transesterification reaction takes place over a set period. Once the reaction is complete, the products are settled and separated. Although simple and effective for variable feedstocks, batch processing can be time-consuming and labor-intensive, which may limit productivity.

For large-scale, commercial biodiesel production, continuous-flow reactors are preferred. These systems allow a steady stream of reactants to enter the reactor while products are continuously removed, enabling higher throughput and greater efficiency. Continuous processes are more suitable for standardized, consistent feedstock and benefit from automation, reduced downtime, and improved heat and mass transfer. After the transesterification reaction, the resulting mixture undergoes separation and purification (Bernardo et al., 2020). The two main products—biodiesel (methyl esters) and glycerol—are separated based on their density differences. Glycerol, being denser, settles at the bottom and is drained off. The crude biodiesel typically contains residual methanol, catalyst, soap, and small amounts of water, which must be removed through washing, drying, and polishing stages. These purification units are essential to meet international fuel standards (such as ASTM D6751 or EN 14214) and to ensure the fuel's compatibility with diesel engines. Proper purification also enhances storage stability and reduces engine wear, making the biodiesel a reliable alternative to petroleum diesel.

5.4 Quality control and testing of biodiesel

Ensuring biodiesel meets quality standards is crucial for engine performance and market acceptance. Key international standards include ASTM D6751 (USA) and EN 14214 (Europe).

5.4.1 Key quality parameters

Quality control is a critical component in biodiesel production, ensuring that the final product meets established fuel standards and is suitable for engine use. Several key parameters are routinely measured and controlled to guarantee performance, safety, and regulatory compliance. One of the primary quality indicators is ester content, which must exceed 96.5% according to the European EN 14214 standard. High ester content ensures that the conversion of triglycerides into biodiesel is complete and that the fuel possesses the desired combustion characteristics. Incomplete conversion results in residual mono-, di-, or triglycerides, which can lead to engine deposits and clogging (Bruhwiler et al., 2021).

Viscosity is another important property, directly affecting the fuel's flow behavior and atomization during injection. Biodiesel must maintain a viscosity within the range of 3.5–5.0 mm^2/s at 40°C. Fuels that are too viscous may not atomize properly, reducing combustion efficiency, while overly thin fuels can lead to leakage and inadequate lubrication. The acid value reflects the presence of FFA in the biodiesel. Elevated acid values can result in corrosion of fuel system components and injectors, especially in modern engines. A low acid value is essential for long-term engine compatibility and storage stability (Bumharter et al., 2023).

Flash point, the temperature at which vapors ignite, is a vital safety measure in handling and transport. Biodiesel typically has a flash point above 120°C, which makes it much safer than petroleum diesel. A lower flash point may indicate methanol contamination, which poses safety and performance risks. Water and sediment content must be minimized because they can cause fuel filter plugging, microbial growth, and corrosion. These contaminants can also disrupt the combustion process, leading to incomplete burning and increased emissions. Finally, sulfur content must be kept extremely low (Cathcart et al., 2021). High sulfur levels contribute to the formation of sulfur oxides during combustion, which are harmful pollutants. Stringent limits (often below 10 ppm) are imposed to align with emission regulations and to ensure compatibility with modern exhaust treatment systems. Together, these parameters define biodiesel's quality and ensure that it performs reliably in diesel engines without causing operational or environmental problems. Regular testing and adherence to standards such as ASTM D6751 (U.S.) or EN 14214 (EU) are essential for maintaining product integrity in commercial production.

5.4.2 Analytical methods

To ensure the biodiesel produced is of high quality and meets international standards, several analytical techniques are employed during the testing phase. These techniques allow producers to monitor chemical composition, detect impurities, and evaluate physical properties critical to engine performance and fuel stability.

Gas chromatography (GC) is the primary method used to determine ester content and residual methanol in biodiesel. It provides precise quantification of FAMEs, which are the active components of biodiesel, and detects any unreacted methanol that could lower the fuel's flash point and pose safety risks. The GC process separates the components of the biodiesel sample based on their volatility, offering a clear profile of its composition. Titration is a classical chemical technique used to determine the acid value of biodiesel (Chen et al., 2021). This test measures the amount of potassium hydroxide (KOH) needed to neutralize the FFAs in a sample. A high acid value indicates incomplete conversion or feedstock issues and suggests that the fuel may be corrosive or unstable during storage.

To assess water content, Karl Fischer titration is the gold standard. This method is highly sensitive and accurate, capable of detecting even trace amounts of water. Water in biodiesel can cause microbial growth, fuel system corrosion, and poor combustion, making this test essential for ensuring both performance and shelf life. Finally, the Cold Soak Filtration Test (CSFT) evaluates the cold flow properties of biodiesel (Chen et al., 2022). This test determines whether the biodiesel will form waxy precipitates or gels under low temperatures, which could clog fuel filters and lines. Biodiesel with poor cold flow properties can be problematic in colder climates, so CSFT is crucial for assessing the fuel's suitability in different environmental conditions. Together, these quality testing methods ensure that biodiesel meets safety, efficiency, and environmental standards, protecting both engines and the broader fuel supply chain. Regular and accurate testing is therefore integral to any successful biodiesel production operation.

5.4.3 Postprocessing steps

Following the transesterification process in biodiesel production, postreaction purification is a critical step to ensure that the final biodiesel product meets fuel quality standards and is safe for engine use. The first stage in this purification process is glycerol separation, as glycerol is a major by-product of transesterification. Because glycerol has a significantly higher density than biodiesel, it can be effectively separated through decantation, where it settles at the bottom of the reactor and is drained off. In larger or more advanced setups, centrifugation is employed for faster and more complete separation, increasing process efficiency.

After glycerol is removed, the biodiesel still contains residual catalysts, unreacted methanol, soaps, and other impurities. To eliminate these, the biodiesel undergoes a washing process. This typically involves gently mixing the crude biodiesel with warm water to extract soluble contaminants. The washing step may be repeated multiple times until the water runs clear, indicating the removal of residual impurities. Care is taken during washing to avoid emulsification, which can trap water and slow down the purification process. The final step is drying, which is essential to remove any moisture absorbed during the washing stage or introduced during earlier processing. Moisture in biodiesel can lead to microbial growth, corrosion of fuel system components, and poor combustion performance (Chojnacka and Moustakas, 2024). Drying can be achieved using heat, vacuum drying, or air bubbling methods depending on the scale of production and available technology. These purification steps: glycerol separation, washing, and drying: are fundamental to producing high-quality, stable biodiesel. They ensure that the fuel is free from contaminants that could affect engine performance or durability and that it meets established specifications such as EN 14214 or ASTM D6751.

5.5 Environmental and economic impacts

Using agricultural residues for biodiesel production presents a sustainable and economically viable alternative to fossil fuels, aligning with broader goals of environmental protection and rural development (Cortez et al., 2022). Unlike first-generation biofuels that often rely on edible oils and compete with food crops, residue-based biodiesel leverages nonfood biomass, such as crop waste, used cooking oil, and agro-industrial by-products (de Bruijn et al., 2006a). This approach not only reduces the pressure on arable land and food supply chains, but also supports circular economy models, where agricultural waste is efficiently repurposed into valuable energy products.

From an environmental perspective, biodiesel derived from residues significantly lowers lifecycle greenhouse gas emissions: with reductions ranging from 60% to 90% compared to petroleum diesel. This makes it a key contributor to climate change mitigation strategies. Furthermore, the process promotes waste valorization, turning what was once a disposal issue into an economic resource (de Bruijn et al., 2006b). By using materials that would otherwise be burned, dumped, or left to decay, the biodiesel industry contributes to waste management and soil conservation. Economically, residue-based biodiesel production enhances rural development. It creates new revenue streams for farmers, especially those in regions with abundant agricultural waste such as rice husk, cotton stalks, or oilseed residues. In addition to providing supplementary income, it generates local employment through collection, processing, and distribution activities. This decentralized model of production ensures that value addition happens closer to the source, reducing urban migration and supporting regional energy security.

5.6 Challenges and future directions

- Feedstock variability: Inconsistent oil content and availability
- Technological constraints: Need for affordable, scalable, and modular systems
- Cost competitiveness: Dependent on oil prices and subsidies
- Policy support: Needed for market stability and investment

5.6.1 Future innovations include

- Catalyst recycling
- Integrated biorefineries
- Enzyme technology for mild, green processing
- Advanced feedstock logistics and preprocessing systems

References

Babin, A., Vaneeckhaute, C., Iliuta, M.C., 2021. Potential and challenges of bioenergy with carbon capture and storage as a carbon-negative energy source: a review. Biomass Bioenergy. 09619534. 146 (105968). doi:10.1016/j.biombioe.2021.105968.

Bachanek, K.H., Tundys, B., Wiśniewski, T., Puzio, E., Maroušková, A., 2021. Intelligent street lighting in a smart city concepts—a direction to energy saving in cities: an overview and case study. Energies. 1996-1073. 14 (11), 3018. doi:10.3390/en14113018.

Bakkaloglu, S., Hawkes, A., 2024. A comparative study of biogas and biomethane with natural gas and hydrogen alternatives. Energy Environ. Sci. 1754-5692. 17 (4), 1482–1496. doi:10.1039/d3ee02516k.

Balcioglu, G., Jeswani, H.K., Azapagic, A., 2022. Evaluating the environmental and economic sustainability of energy from anaerobic digestion of different feedstocks in Turkey. Sustain. Prod. Consum. 23525509. 32, 924–941. Available from: http://www.journals.elsevier.com/sustainable-production-and-consumption/. doi:10.1016/j.spc.2022.06.011.

Barr, M.R., Volpe, R., Kandiyoti, R., 2021. Liquid biofuels from food crops in transportation—a balance sheet of outcomes. Chem. Eng. Sci. X. 25901400. (10), 100090. doi:10.1016/j.cesx.2021.100090.

Bernardo, G., Araújo, T., da Silva Lopes, T., Sousa, J., Mendes, A., 2020. Recent advances in membrane technologies for hydrogen purification. Int. J. Hydrogen Energy. 03603199. 45 (12), 7313–7338. doi:10.1016/j.ijhydene.2019.06.162.

Bruhwiler, L., Basu, S., Butler, J.H., Chatterjee, A., Dlugokencky, E., Kenney, M.A., McComiskey, A., Montzka, S.A., Stanitski, D., 2021. Observations of greenhouse gases as climate indicators. Clim. Change. 15731480. 165 (1-2). Available from: https://www.wkap.nl/journalhome.htm/0165-0009. doi:10.1007/s10584-021-03001-7.

Bumharter, C., Bolonio, D., Amez, I., Jesús García Martínez, M., Ortega, M.F., 2023. New opportunities for the European Biogas industry: a review on current installation development, production potentials and yield improvements for manure and agricultural waste mixtures. J. Clean. Prod. 09596526. 388, 135867. doi:10.1016/j.jclepro.2023.135867.

Cathcart, A., Smyth, B.M., Lyons, G., Murray, S.T., Rooney, D., Johnston, C.R., 2021. An economic analysis of anaerobic digestate fuel pellet production: can digestate fuel pellets add value to existing operations? Clean. Eng. Technol. 26667908. 3. Available from: https://www.sciencedirect.com/science/journal/26667908. doi:10.1016/j.clet.2021.100098.

Chen, B., Shao, Y., Shi, M., Long, J., Qingyao, H., Yan, S., 2021. Anaerobic digestion of chicken manure coupled with ammonia recovery by vacuum-assisted gas-permeable membrane process. Biochem. Eng. J. 1369703X. (175), 108135. doi:10.1016/j.bej.2021.108135.

Chen, L., Msigwa, G., Yang, M., Osman, A.I., Fawzy, S., Rooney, D.W., Yap, P.-S., 2022. Strategies to achieve a carbon neutral society: a review. Environ. Chem. Lett. 1610-3653. 20 (4), 2277–2310. doi:10.1007/s10311-022-01435-8.

Chojnacka, K., Moustakas, K., 2024. Anaerobic digestate management for carbon neutrality and fertilizer use: a review of current practices and future opportunities. Biomass Bioenergy. 09619534. 180, 106991. doi:10.1016/j.biombioe.2023.106991.

Cortez, S.C., Cherri, A.C., Jugend, D., Jesus, G.M.K., Bezerra, B.S., 2022. How can biodigesters help drive the circular economy? An analysis based on the SWOT matrix and case studies. Sustainability. 2071-1050. 14 (13), 7972. doi:10.3390/su14137972.

de Bruijn, H., van Duin, R., Huijbregts, M.A.J., Guinee, J.B., Gorree, M., Heijungs, R., Huppes, G., Kleijn, R., de Koning, A., van Oers, L., Sleeswijk, A.W., Suh, S., Udo de Haes, H.A., 2006a. Impact Assessment. Springer Science and Business Media LLC, pp. 63–95. doi:10.1007/0-306-48055-7_9.

de Bruijn, H., van Duin, R., Huijbregts, M.A.J., Guinee, J.B., Gorree, M., Heijungs, R., Huppes, G., Kleijn, R., de Koning, A., van Oers, L., Sleeswijk, A.W., Suh, S., Udo de Haes, H.A., 2006b. Interpretation. Springer Science and Business Media LLC, pp. 97–108. doi:10.1007/0-306-48055-7_10.

CHAPTER 6

Pyrolysis and gasification of agricultural residues

Balaji Doolam, Matta Chenna Keshava Charan, Bishwambhar Mishra, Rajasri Yadavalli, Sanjeeb Kumar Mandal, Sumithra Salla, and C. Nagendranatha Reddy
Department of Biotechnology, Chaitanya Bharathi Institute of Technology, Hyderabad, Telangana, India

6.1 Introduction

"The population of the earth is predicted to rise to 9.8 billion people by the year 2050" (UN, 2022). Due to this growth, the need for food is escalating, placing stress on the agricultural industry. Climate change and global warming have made agriculture more challenging by causing resource depletion and poor management. Currently, major energy sources such as coal and petroleum provide 80% of India's energy needs. Long-term usage of petroleum fuels, however, has adverse consequences on the environment, including emissions of carbon dioxide and other pollutants. India's dependence on energy imports raises questions about energy security because of the uncertainty it causes in terms of price and supply. India has been importing a lot more crude oil over the years. Given that, over the span of 25 years, India's consumption of energy is expected to increase 3–4 times, it is imperative to investigate more ecologically friendly, sustainable, and cleaner energy sources (Tagade et al., 2021; Venkata Mohan et al., 2016).

With the International Energy Agency projecting a four-fold growth in demand, the bioenergy sector is being highlighted as a critical countermeasure for a low-carbon future (Röder et al., 2020). Due to its agrarian economy, India has a substantial potential for agro-waste biomass, approximately 500 metric tons annually (Bhuvaneshwari et al., 2019). In developing nations such as India, agro-residues have drawn attention as a possible source of bioenergy, especially considering India's heavy reliance on agriculture (Vivek et al., 2019; Nagendranatha Reddy et al., 2022; Mishra et al., 2023).

By 2050, biomass is to account for roughly half of global energy use, with industrialized nations using it up to 80% of the world's total (Osman

Bioenergy from Agricultural Residues
https://doi.org/10.1016/B978-0-443-24726-2.00009-2

et al., 2021). Climate change, animal welfare, the environment, and human health are all significantly impacted by agricultural production. It is essential to cut greenhouse gas emissions if we want to avoid a temperature increase of at least 35.6°F (Woods et al., 2010). Proper disposal of agro-residues is a major concern because open-burning techniques are still commonly used, which has detrimental effects on soil and air quality and leaves much of the agro-waste unused (Makavana et al., 2020). Uncontrolled burning of agricultural solid wastes releases gases that have an impact on the climate and hinder food production. The proper disposal of agricultural solid wastes has several positive consequences, such as reduced food production and reduced greenhouse gas emissions.

Various pathways, including biochemical and thermochemical methods, can convert agricultural waste biomass into bioenergy. The horticultural industry faces the challenge of meeting the food demand of a growing global population while minimizing the use of agrochemicals due to their detrimental impacts (Sharma et al., 2022a, 2022b).

Environmental protection, resource conservation, soil health, climate mitigation, and economic gains all depend on sustainable waste management in agriculture. It is possible to turn biomass into energy utilizing thermochemical and biochemical processes (Bhaskar et al., 2020). Biochemical techniques such as anaerobic digestion and fermentation are used to convert cellulose and hemicellulose into biofuel during the hydrolysis and fermentation stages. Currently, biochemical approaches are expensive and have issues using biomass that is high in lignin (Zhang et al., 2017). In addition, these techniques are unstable in nature. They are characterized by very moderate speeds, and the production process uses a lot of water that is circulated, diluting the final product (Sikarwar et al., 2016; Li et al., 2021).

Agricultural residues (ARs) are primarily composed of lignocellulosic biomass, and a variety of methods have been developed to convert them into one or more products with added value (Reddy et al., 2023). Fig. 6.1 shows the most widely used methods for turning valuable products from lignocellulosic biomass (agro-residues). To successfully use agro-residues, the two main conversion technologies are thermochemical and biological processes (Saifuddin et al., 2017). Because of many disadvantages associated with microbial and other biological degradation processes, it has been demonstrated that the thermochemical process pathway is more feasible and might be a superior choice for recovering the chemical and energy values of lignocellulosic biomass faster and at a lower cost, especially when

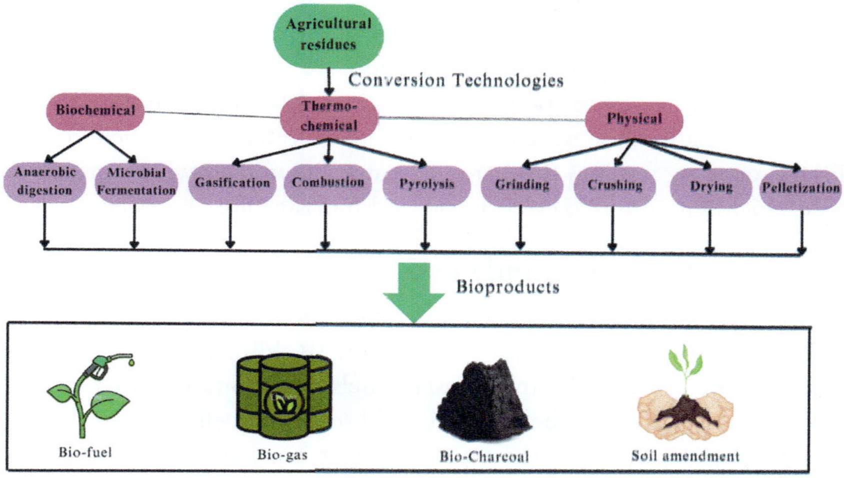

Fig. 6.1 Technologies used to turn lignocellulosic biomass (agro-wastes) into useful products.

it comes to the value-adding of agricultural leftovers (Himmel et al., 2007; Asomaning et al., 2018).

Combustion (Choe, 2021), gasification (You et al., 2018), and pyrolysis (Ahmed et al., 2020) are three thermochemical conversion methods that enable the conversion of material into usable energy. When the chemical bonds of organic compounds are broken in an inert environment between 300°C and 650°C, the procedure is known as pyrolysis. It transforms lignocellulosic biomass into energy and products with additional value. Large biomass molecules are initially broken down by heating, producing solid biochar and bio-oil. Both the environment and the economy profit from this process (Andrade et al., 2018). With no risk to investors, biomass combustion is a commonly used commercial method for generating heat and electricity. The end product, thermal energy, has the ability to produce electricity. The utilization of biomass combustion technologies offers an economic and competitive edge when waste is used as the primary fuel (Glushkov et al., 2021). The process of partially oxidizing biomass to create a mixture of combustible gases is known as gasification. Temperatures between 700 and 900°C are ideal for it to function normally (You et al., 2018).

In the present chapter, we will discuss various types of ARs, the influencing factors, the development of value-added products, and the commercialization of pyrolysis and gasification processes.

6.2 Types of ARs

The byproducts of agricultural production are known as ARs. They fall into two primary categories: Materials such as stalks, stems, leaves, and seed pods that remain in the field after harvesting are referred to as field residues (FR). Processing residues (PR) are the materials left over when agricultural goods, including husks, seeds, bagasse, and molasses, are processed.

6.2.1 Rice straw, bran, and husks

Worldwide, rice is grown to a considerable extent and is thought to produce about 618 million tons of rice annually (Rahimi-Ajdadi et al., 2018). For the manufacture of biomaterials, biofuels, and biomolecules, rice straw (RS), which is an FR, is the most plentiful source of cellulose, yielding 1.5 tons of waste for each metric ton of rice that is produced. Other residues from agriculture are bran, husk (a PR), and straw (an FR) (Athira et al., 2019). The principle elements of RS are cellulose (35%), hemicellulose (27%), lignin (24%), and ash (14%) (Guzmán et al., 2015).

6.2.2 Wheat straw and bran

Over one-third of the world's population consumes wheat, the second most significant food grain after rice (Shahryari et al., 2018). It is expected that 12.88 million tons will be produced annually. The wheat straw (WS) is mainly composed of cellulose, hemicellulose, and lignin, which also contain vitamins and bioactive substances (Khan and Mubeen, 2012). The WS comes under the FR category, whereas wheat bran (WB) is classified as PR.

6.2.3 Maize stalk and husk

An important plant, maize, produces 875,226,630 tons (Ranum et al., 2014) with a 71% carbohydrate content worldwide. It produces a lot of waste, such as straw (an FR), husks (a PR), corncobs, and chaff. By-product maize straw (an FR) comprises 16.7%–18.9% lignin, 25.7%–25.4% hemicelluloses, and 28.4%–37.5% cellulose (Guo et al., 2015).

6.2.4 Cassava stalks and peelings

With an efficiency of 10.95 t/ha and a global yield of 202.65 million tons, cassava is a significant food crop. Because of its 27.92 t/ha highest output rate, India's production of 6.7 million tons, produced on 240,000 ha, is noteworthy (Edison et al., 2006). The cassava root, which makes up 11% of the root, is processed to produce byproducts such as cassava flour, cellulose, hemicellulose, and lignin. The cassava peel, which is processed to yield byproducts high in carbohydrates (Barati et al., 2020). The cassava stalks come under the FR category, whereas peelings are classified as PR.

6.2.5 Groundnut shells and straw

With a 38.2 million ton annual yield, groundnut is a leguminous plant used for its seed and oil. With 25%–30% of the pod's weight made up by the shell, or outer layer, the largest remnant, the seed makes up the remainder (Perea-Moreno et al., 2018). The groundnut shells come under the FR category, whereas groundnut straw is classified as PR.

6.2.6 Sugarcane bagasse (SB)

Globally, 1.6 billion tons of sugarcane are grown annually and are used to manufacture ethanol and sugar (Jugwanth et al., 2020). After crushing and extracting the juice, sugarcane bagasse (SB) (a PR), which contains 30% and 60%–80% carbohydrates, is produced. Its components include polysaccharides, lignin (22%), hemicellulose (28%–30%), and cellulose (33%–66%) (Sidana and Farooq, 2014).

6.2.7 Soybean pods and straw

A significant crop, soybean is used as both food and fodder by people and is produced in 312 million metric tons annually (De Pretto et al., 2018). After the seeds are harvested, stems, leaves, and pods (an FR) are produced to make soybean straw (a PR) and soybean pod husk (a PR). Similar to WS in terms of its high nitrogen concentration, a cheap, sustainable energy source is soybean straw and chemicals. It contains protein and pectin and is used to feed mixed-breed cows. Its constituents are cellulose (35%–83%), insoluble lignin (14%–21%), hemicelluloses (17%), ash (11%), and acid-soluble lignin (1%) (Ginni et al., 2021).

6.3 Pyrolysis: A waste-to-energy technique

A number of different approaches have been investigated to turn biomass into energy in the search for sustainable energy sources. Gasification, pyrolysis, and combustion are a few of these techniques. Every technique has distinct features and procedures for transforming biomass into energy or beneficial byproducts. The thermochemical process of pyrolysis is the breakdown of organic molecules (such as biomass) in the absence or with a restricted amount of oxygen. Usually, this procedure takes place at high temperatures of 400°C to 800°C. When biomass is heated under controlled conditions, a process known as pyrolysis, it breaks down into three primary products: bio-oil, biochar, and syngas, which is a mixture of gases primarily H and CO.

The promise of pyrolysis as a sustainable and effective approach for biomass conversion has been highlighted by studies such as those conducted by Haris et al. (2021), Saikia et al. (2018), and Ahmed et al. (2020). Their study highlights how feasible it is to produce energy with as little negative impact on the environment as possible.

6.3.1 Types of pyrolysis processes

There are five classes for biomass conversion: slow, fast, flash, vacuum, and hydro-pyrolysis (HP). according to the pyrolysis processing parameters. The different pyrolysis processes are illustrated in Fig. 6.2.

6.3.1.1 *Slow pyrolysis (SP)*

At pyrolysis temperatures ranging from 300°C to 800°C, slow pyrolysis (SP) decreases the bio-oil yields, which encourages the synthesis of biochar (Ahmad et al., 2014). Most of the literature focuses on the synthesis of

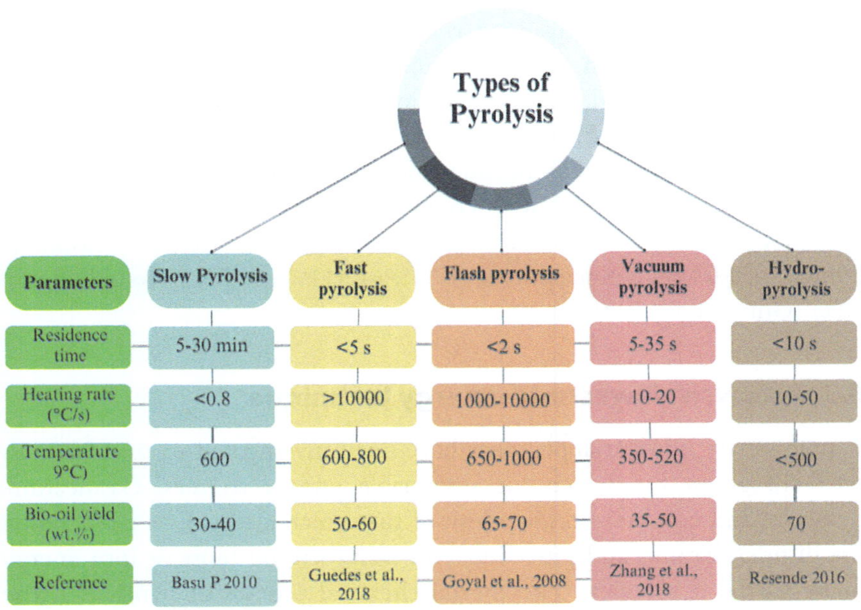

Fig. 6.2 Various types of pyrolysis processes.

biochar; however, SP is also utilized to produce compounds such as acetic acid, furfural, and phenols (Hagner et al., 2020). When SP is used and biochar is the target outcome, carbonization is a biomass conversion technology for producing charcoal with an optimal feedstock moisture content of 15%–20% (Eke et al., 2020).

6.3.1.2 Fast pyrolysis (FP)

According to Roy and Dias (2017), with a heating rate of up to 1000°C/s and a residence time of <2 s, the temperature range for fast pyrolysis's (FP's) response is 400°C–1000°C. According to Pecha and Garcia-Perez (2020), the FP method is distinguished by tiny particle size, fast heating rate, and temperatures (400 and 650°C) (Anca-Couce, 2016). To produce biofuel and chemicals, FP is a versatile and favored technique for turning biomass into a liquid that is simple to transport and store (Li et al., 2022).

6.3.1.3 Flash pyrolysis (FLP)

Flash pyrolysis (FLP) requires three major conditions, viz., high temperatures (900°C–1300°C), a heating rate (>1000°C/s), and dwell/residence time (<0.5 s). The product contains >75% bio-oil. To achieve efficient heat transmission, the biomass must be ground into a fine powder and pretreated (105–250 m) prior to pyrolysis (Reza et al., 2023a). With the right conditions, FLP can deliver 60–75 wt percent of oil. Variations in these major conditions might influence the final product's quality and quantity. FLP oil with a high water content (>15 wt%) has a greater value of heating of 15–36 MJ/kg. It is quicker and has a higher feedstock handling capacity while maintaining a similar reactor volume and processing complexity. In contrast to other pyrolysis methods that mostly produce solid and gaseous byproducts, FLP produces liquid bio-oil as its main end product (Ighalo et al., 2022).

6.3.1.4 Vacuum pyrolysis (VP)

Utilizing vaccum pyrolysis (VP), feedstock can be thermally broken down in a reactor operating at subatmospheric pressure without the presence of oxygen. According to Gabhane et al. (2020), the ranges for pressure, temperature, and heating rate are 0.01–0.20 MPa, 300–700°C, and 0.1–1°C/s, respectively. The VP reaction produces high bio-oil yields because secondary degradation, which is necessary to produce biochar, is suppressed. This is because higher temperatures lead to a greater removal of VM, which causes increased heating and biomass decomposition. The yield of biochar is compatible with other pyrolysis techniques; however, it is reduced when the temperature is raised (Liu et al., 2018).

6.3.1.5 Hydro-pyrolysis

The HP method uses a high-pressure hydrogen atmosphere, which has the following properties: temperature of pyrolysis (350–600°C), heating rate (10–300°C/s), pressure (10–17 MPa), and residence length (>60 s) (Oh et al., 2021). Wang and Song (2018) found that coloading Zn and Ga in HP enhances the yield of aromatic hydrocarbons by 37.4%. But the bio-oil's oxygenated molecules make it unsuitable as a fuel for automobiles, necessitating additional enhancements SUCH AS hydrotreating; it raises the cost and complexity of the procedure and reduces the appeal of HP as a standalone method (Kong et al., 2020).

6.3.2 Factors affecting the pyrolysis

The composition and yield of different compounds from pyrolysis are influenced by multiple parameters, such as reactor design, temperature, gas flow rate, heating rate, and particle size (Akhtar et al., 2012). By optimizing these factors, the production and quality of bio-oil, biogas, or biochar can be increased; the following are the effects of operational circumstances on these products.

6.3.2.1 Temperature

The productivity and efficiency of pyrolysis processes and the final products are greatly influenced by temperature. The best conditions for producing bio-oil are between 400 and 550°C. Due to secondary cracking processes including decarboxylation, deoxygenation, and dehydrogenation that turn tars and vapors into noncondensable gases at higher temperatures, production is decreased (Uddin et al., 2018). The study conducted by Colantoni et al. (2016) explored the generation of biochar from pelletized agricultural waste, specifically grapevine and sunflower husk, through pyrolysis at temperatures of 400 and 500°C. The results indicated that the yield of solid biochar decreased as the temperature increased. Biochar is a porous and carbon-rich material that has the ability to improve soil quality and serve as a long-term environmental energy storage solution.

The highest bio-oil production rates were found to be 47.3%, 36.7%, 28.4%, and 38.1% for corn cob, WS, RS, and rice husk, respectively, in pyrolysis tests. Biomass type and pyrolysis temperature significantly influenced the composition and yield of pyrolysis products (Biswas et al., 2017).

6.3.2.2 Particle size

During pyrolysis, the mass and heat transmission rates are influenced by the biomass particle's size, which in turn influences the bio-oil output. High bio-oil yields are favored by the quick decomposition of small particles, but the generation of char is favored by the slow decomposition of larger particles (Tripathi et al., 2016). Gas yield decreases with increasing particle size, whereas yields of tar and char increase because of their decreased surface area and increased thermal resistance. Particle size also affects product yield. Due to their rapid disintegration and involvement in secondary processes, biomass particles smaller than 0.6 mm improve biogas yield while decreasing biooil yield (Varma and Mondal, 2017).

Hong et al. (2020) examined the effects of pyrolysis conditions on two distinct crop residues (cotton stalk and RS) at varied temperatures and particle sizes. The samples with the highest yield, H, O, and carbon conversion rate among the various particle sizes were those with coarse grinding (particle size <10 mm), while the samples with the highest percentages of CO and CH_4 as well as higher energy conversion rates of gases and heating values were those with fine grinding (particle size <300 μm). The sample that was ground to an ultrafine particle size (<50 μm) had the smallest pore size; the highest pH, C, specific surface area, pore volume, and functional group strengths; but the lowest charcoal and gas yields.

Sahoo et al. (2021) investigated the pyrolysis of ARs using RS, WS, and SB in a fixed-bed reactor. Because large particles do not promote mass transfer or heat transfer due to the breaking process and uneven heating of the particle, the yield of bio-oil decreased when the size of the particle increased (0.2–0.8 mm).

6.3.2.3 Gas flow rate

Due to side reactions including thermal cracking, repolymerization, and biochar recondensation caused by the pyrolysis process, which converts biomass and generates a significant volume of vapor, the amount of biooil produced is reduced (Akhtar and Amin, 2012). Due to its stability in chemicals, accessibility, and quantity, nitrogen gas is a favored way for removing vapors from pyrolysis reactors. In addition, the gas's higher flow rate shortens the period that vapors remain in the reactor (Tripathi et al., 2016).

At a temperature of 550°C, with a particle size ranging from 0.2 to 0.4 mm and a constant heating rate of 100°C/min, the impact of varying sweeping gas flow rates (ranging from 20 to 100 mL/min) on the distribution of products derived from SB was evaluated. The highest bio-oil yield was observed at a flow rate of 30 mL/min, followed by a rapid decline. At lower flow rates, the prolonged residence time of released volatiles in the reactor led to increased vapor-phase secondary cracking and repolymerization processes, ultimately resulting in higher char and gas production, accompanied by a reduction in liquid yield (Sahoo et al., 2021).

6.3.2.4 Heating rate

The rate of heating has a major impact on the pyrolysis process's yield since it accelerates the breakdown of biomass vapors, which produces more gases than charcoal (Solís Hontañón, 2016). Chen et al.'s (2017) study investigated the thermal conversion properties, kinetics, and thermodynamics of rape straw and WB under nonisothermal conditions. The research highlighted significant differences in the pyrolysis properties of RS and WB. Notably, increasing the heating rates from 10 to 30 K/min greatly improved the performance of RS and WB during pyrolysis.

In the investigation conducted by Sahoo et al. (2021), the pyrolysis of agricultural residues, including RS, WS, and SB, was examined in a fixed-bed reactor. As the heating rate increased from 20 to 100°C/min, there was an increase in bio-oil yield for SB, WS, and RS, reaching 52.5% (from 47.2%), 37.2% (from 31.9%), and 32.6% (from 27.3%), respectively. Following a similar trend, the biochar yield decreased to 32.6%, 38.9%, and 40.1% for SB, WS, and RS, respectively.

6.3.2.5 Reactors

Pyrolysis has been carried out in reactors with fixed beds, fluidized beds (FBs), rotative, vacuum, plasma, and microwave designs. Another popular choice is the microwave reactor, which produces 60–70% bio-oil due to its rapid heating rates, high temperatures, and short residence durations (Bhoi et al., 2020). However, its usage is restricted by the requirement for microwave absorbers, expensive processing, and high power consumption (Asomaning et al., 2018).

SP in a fixed-bed reactor is an easy and inexpensive way to produce biochar. Charcoal is the primary product after the biomass has been burned in the reactor for several hours or days. This method is simple to use and simply requires a chimney (Bamido, 2018).

The screw reactor, also known as the auger reactor, is a continuous pyrolysis technology that transports feedstock through a heated reactor using a rotating helical screw as a conveyor. This method ensures uniform mixing and boosts heat transfer efficiency, making it appropriate for biomass of different sizes and shapes (Campuzano et al., 2018).

One of the main study areas for microwave reactors, which use microwave heat to transmit energy to molecules or atoms, is the pyrolysis procedure. By using inert gas as the carrier gas, these reactors, which are frequently utilized in microwave cavity ovens, maintain an atmosphere free of oxygen. Microwave reactors are effective at transmitting heat, have improved chemical reactivity, and have control over the rate of heating, which makes them useful at producing valuable chemicals from biomass. They can produce large amounts of syngas and biooil from a variety of biomasses and industrial wastes (Reza et al., 2023a).

In the study conducted by Sahoo et al. (2021), RS, WS, and SB subjected to pyrolysis in a fixed-bed reactor revealed a significant increase in the production of bio-oil by using small particles and a high heating rate. Furthermore, the analysis of noncondensable gases showed the coexistence of H_2, CO, CO_2, and CH_4 molecules. Higher concentrations of H_2 and CH_4 in these gases were found to enhance the calorific value of the produced gases.

In summary, this section discusses numerous important parameters that impact the pyrolysis products' production and makeup, including charcoal, biogas, and bio-oil. Particle size, temperature, and gas flow rate, heating rate, and the kind of reactor being used are all included in this list. Understanding and optimizing these parameters are crucial for achieving desirable outcomes in biomass pyrolysis processes and utilizing all of the possibilities of these valuable products for various scientific and industrial applications.

6.3.3 Value-added products from pyrolysis

The intricate process of biomass pyrolysis is influenced by many factors, such as biomass feedstock, reaction parameters, primary component reactions, and other factors. Three main byproducts are produced during the pyrolysis of biomass: pyrolytic gas, char, and bio-oil.

6.3.3.1 Bio-oil

Bio-oil, with molecular weights ranging from 370 to 1000 g/mol and mostly impacted by biomass sources, is a mixture of water and solid particles (Wang et al., 2017). This process involves a variety of chemicals, including organic acids, alcohols, aldehydes, ketones, phenols, esters, ethers, furans, nitrogen, and sulfur compounds (Milne et al., 1997; Anca-Couce A 2016). Bio-oils are created during pyrolysis and are viscous/nonviscous, dark brown, and strongly fragrant. They are created by the thermal breakdown of cellulose and can be used as fuel for motors, boilers, and furnaces. Bio-oils are thought of as energy-efficient, environmentally friendly, and appropriate for rural socioeconomic issues. They can also be utilized to synthesize resins, adhesives, lubricants, flavorings for wood, and preservatives (Reza et al., 2023a).

The highest bio-oil production rates were found to be for corn cob (47.3%), WS (36.7%), RS (28.4%), and rice husk (38.1%) in pyrolysis tests. Biomass type and pyrolysis temperature significantly influenced both pyrolysis product yield and composition (Biswas et al., 2017).

6.3.3.2 Biochar

Biochar is a carbonaceous result of biomass pyrolysis made up of unconverted organic solids and carbonaceous byproducts of the partial or total breakdown of biomass components (Ahmad et al., 2014). The increased ash content brought on by inorganic materials facilitates the synthesis of biochar (Luo et al., 2023). The carbon content of char can be between 53 and 96 wt percent, and its yield and heating value can be between 30 and 90 wt percent and 20 and 36 MJ/kg, respectively (Mohan et al., 2014). It can be used as fertilizer, to clean up pollutants, to recycle agricultural waste and byproducts, to catalyze reactions, to store energy, and to protect the environment (Ahmad et al., 2014). Recently, the process of making activated carbon, one of the most effective adsorbents for a variety of specific uses, has shown great potential using biochar (Reza et al., 2022). Biochar was proposed as a potential replacement for electrodes, cathode catalysts, and proton exchange membranes in microbial fuel cell (MFC) applications by Chakraborty et al. (2020). Using wastewater as a substrate, MFC is able to transform the energy of organic materials found in chemical bonds into electrical energy. Biochar shows promise as a cathode catalyst and as an electrode material in MFC (Nagendranatha Reddy et al., 2020).

Liu et al. (2018), studied the physicochemical characteristics of biochars derived from various straw sources, including cotton, rapeseed, corn, wheat, and RSs, which were investigated. Higher pyrolysis temperatures were found to slightly increase carbon content, pH, and electrical conductivity while reducing the O/C and H/C ratios. In addition, as the carbonization temperature increased, the presence of oxygen functional groups decreased. The carbonization temperature and crop residue type both have an effect on combustion performance. Notably, cotton stalk biochar exhibited the most favorable attributes for solid biofuel generation due to its lowest H/C ratio and highest fixed carbon content.

6.3.3.3 Pyrolytic gas

Depending on the fuel, pyrolysis generates gases known as syngas, biogas, or bio-syngas. As a result of secondary vapor splitting, gas output rises as temperature rises (Reza et al., 2023b). CO_2, CO, H_2, low carbon number hydrocarbons such as CH_4, C_2H_6, and C_2H_6, in addition to minute amounts of other gases including C_3H_8, NH_3, nitrogen oxides, sulfur oxides, and low carbon number alcohols, are all produced during the pyrolysis of biomass (Qu et al., 2011; Reza et al., 2023b). For instance, syngas can be utilized directly as fuel in microtubular solid oxide fuel cells (Nagendranatha Reddy and Min, 2019). In addition, syngas can be used as a fuel for gas turbines that produce electricity, as a heat source for pyrolysis, and as a combustion fuel in industry (Reza et al., 2022).

Patra et al. (2021) investigated the pyrolysis of oat hull in bio-oil production at high temperatures (600°C) to enrich it with acids, alcohols, and phenols. The gaseous product obtained at 600°C yielded more CH_4 and H_2, with a higher heating value of 8.4 MJ/Nm^3.

In conclusion, biomass pyrolysis yields three primary byproducts: bio-oil, biochar, and pyrolytic gas, each with unique characteristics and potential applications in various fields of science and industry.

6.4 Gasification: Converting residues to energy

Gasification is a thermochemical technique that produces syngas, a flammable gas, from organic or carbonaceous materials. Syngas is a combination of gases such as H, CO, and other gases that can be utilized to make fuels, chemicals, energy, and heat. Numerous organic materials, such as coal, ARs, forestry residues (Narnaware and Vahora, 2022), and municipal solid waste, can be processed to create syngas. ARs are extremely abundant and renewable, making them ideal for gasification. Gasification is the process of converting carbonaceous materials into gaseous fuels by using gasifying agents such as air, oxygen, and steam. It seeks to produce the most gaseous content while producing the least amount of unreacted char and condensable hydrocarbons. In order to reduce unreacted chars and condensable hydrocarbons, high temperatures are used to create flammable gas combinations. Alkali and alkaline earth metals and catalysts based on nickel can dramatically lower reaction temperature and time for gas generation. However, gasification also results in the production of tar and fly ash, contaminants in producer gas that can clog turbines, pipelines, and engines (Rahimi et al., 2022).

6.4.1 Factors influencing gasification

There are some major factors that influence the gasification process. They are:

6.4.1.1 Biomass characteristic/feedstock

With changing proportions in both the solid and liquid phases, a complex blend of both organic and inorganic components makes up biomass. For plant design planning, accurate classification of biomass fuel, including agricultural, energy, forestry, and industrial waste, is essential. Two important metrics to consider when assessing the pyrolysis and gasification characteristics of biomass are its cellulose and lignin concentrations (Sharma et al., 2015). The gasification process is affected by the char structure and the cellulose-lignin interaction. Higher cellulose concentration lengthens and raises the peak temperature of syngas, while higher hemicellulose content gives larger (cellulose + hemicellulose)/lignin ratios, with lignin creating more tar (Havilah et al., 2022). Pretreatment, a phase in thermochemical conversion, has an impact on biomass preparation, which is important for gasification. The resistant nature of biomass must be overcome through structural alteration. Enhancing the properties of biomass is crucial for improving energy use efficiency. Pretreatment techniques might be physical, chemical, or biological, with physical treatment being the most popular. The drawback is that there is an additional cost associated with buying a pretreated product that has less key component deterioration (Anukam and Berghel, 2021).

6.4.1.2 Moisture content

The syngas composition and gasification efficiency are greatly impacted by the moisture content of the feedstock. There are two possible states for biomass: free and equilibrium. While bed temperatures remain constant below 15% moisture content, conversion is affected by rising moisture content because water evaporation requires more energy (Wu et al., 2009). The moisture content of biomass plays a crucial role in affecting the rate of biomass consumption, lowering reaction zone temperatures by necessitating energy for fuel evaporation, and impacting both gasifier performance and the quality of end products. Beohar et al. (2012) claim that relative humidity and air temperature have an effect on gasifier performance and end-product quality. High moisture content lowers heating value, net calorific value, and overall productivity of the gasification process. The presence of tar in the gas rises in tandem with the moisture content of biomass. As a result, more air is required to burn biomass, which raises the content of CO_2 and lowers the level of CO. At low moisture levels, CO_2 content ranges between 5% and 10%, but at greater moisture levels, it may rise to 15% or higher (Ramanan et al., 2008).

Naryanto et al. (2020) investigated the tar content in wood pellets using a downdraft gasifier system. As a feedstock, wood pellets with moisture contents ranging from 2 to 6 wt.% were employed. The results suggest that increasing the moisture content marginally reduced the total concentration of tar production on the biomass's nine tar components. Furthermore, the lowest moisture content yielded the most light tar on tar classification.

6.4.1.3 Temperature

Temperature significantly affects the rate of gasification and the efficiency of gasifiers because all processes are reversible. The extremely high temperature in the reaction zone is crucial for the gasification cycle's tar and char concentration, gas production and composition, cold gas efficiency, calorific value, and carbon conversion. Temperature has a dominating influence in gasifiers, which can result in a variety of syngas compositions. According to the literature, temperature differences between various gasifiers can result in varied gas compositions (Rong et al., 2015). In gasification, higher temperatures, between 700°C and 900°C, enhance cracking and reduce the amount of tar. This speed ups the reactions in the oxidation and reduction zones, producing syngas with higher efficiency and less tar. For the production of syngas with a high calorific value and minimal tar content, the oxidation and reduction zone temperatures should range within 800°C–950°C and 650°C–900°C, respectively. Endothermic gasification reactions such as secondary cracking and heavy hydrocarbon reforming are encouraged as the bed temperature rises. These endothermic processes promote H_2 generation. Using burned cashew nut shells as fuel, Ramanan et al. (2008) discovered that the fuel's hydrogen content also had a significant influence. Therefore, utilizing this fuel, the syngas' hydrogen percentage was between 10% and 15%. Since Boudouard reactions (C(s) + CO_2 à2CO), which consume CO_2 and raise CO content, predominate at temperatures above 850°C–900°C, the majority of investigations demonstrate that CO_2 concentration decreases as temperature rises. In addition, tar is more easily destroyed and reformatted at high temperatures, which results in a drop in tar content and a rise in gas output (González et al., 2008).

6.4.1.4 Equivalence ratio (ER)

When it comes to figuring out the composition of the generating gas in FB reactors and fixed/moving bed reactors, the equivalence ratio (ER) is crucial. The ER is computed by comparing the stoichiometric and actual air-fuel ratios. ER chooses the air that is provided for gasification. Due to exothermic processes, an increase in ER raises gas production and gasification temperature. Low ERs cause the bed temperature, gas output, and tar formation to all decrease. The ER value for efficient gasification ranges from 0.19 to 0.43, with 0.3 being the theoretical maximum (Zainal et al., 2002).

6.4.1.5 Particle size

The time needed for gasification and the final outcomes are both greatly influenced by the feed particle's size. The rate of thermal diffusion within the particles reduces as particle size rises, resulting in decreased heating rates. The fuel that has to be converted to gas must have a moderate average particle size to maintain a specific rate of biomass consumption and a suitable pressure drop within the reactor. A lower particle size results in a smaller reactor and better producer gas quality. Fixed-bed gasifiers use particles with sizes between 20 and 80 mm, while FB gasifiers use smaller particles. According to Edrich et al. (1985)'s study, the size of the particles has a significant impact on the rate of gasification. When the size of the wood particles in a fixed-bed gasifier is lowered, the gasification rate increases, thereby resulting in more char and less tar by slowing down the heat-up process.

6.4.1.6 Pressure

In the gasification process, atmospheric and greater pressures are frequently used; depending on how the created syngas will be used, the best pressure should be chosen. Higher pressures are desired if the syngas is utilized to produce methanol or synthetic motor fuels to lower the tar concentration and increase process yield. Atmospheric pressure should be employed to produce burnable gases.

At higher pressures, the equilibrium will shift to the left, favoring the production of CO and H. This is because the products of these reactions (CO_2 and H_2O) are smaller molecules than the reactants (CO and H_2) (Tumuluru, 2017).

6.4.2 Value-added products from gasification

Gasification mostly produces H_2, CO, and a small quantity of CH_4. Furthermore, corrosion-prone substances such as chloride and potassium are eliminated by the high temperature. As a result, producing clean gas is environmentally friendly (Sharma et al., 2022a, 2022b). The important product derived primarily from biomass gasification is syngas, which contains (H_2 + CO + CO_2), natural gas, H_2O, traces of various inert gases such as N_2, and some unburned hydrocarbons with impurities (Ceranic et al., 2016). Tar, such as syngas, is a byproduct of the low-temperature gasification process. To boost syngas concentration and further break down hydrocarbons, higher temperatures should be employed in the processes (Kremling et al., 2017).

Another valuable byproduct of gasifying biomass is hydrogen, which has a lot of potential applications in the creation of clean energy. Many studies have been conducted on the biomass for the possibility of producing H_2 as it is one of the renewable resources that can be used to directly produce hydrogen through a range of thermochemical processes such as pyrolysis and gasification techniques (Sekoai and Daramola, 2015; Suhartono and Azizah, 2006). Compared to other methods such as water-gas shift reaction, steam gasification reaction, supercritical water gasification, and air gasification, steam gasification will produce renewable H_2 at greater yields (Baruah and Baruah, 2014). Higher steam-to-biomass ratios during steam gasification will initially lead to higher H_2 concentrations, but further increases in the ratio will lead to a reduction in the efficiency of this process; therefore, the ratio should be at its ideal. Elevating the temperature should also focus tar reduction (Sharma and Sheth, 2016).

Through the production of fuel-grade gases such as syngas and biochar, biomass gasification technology significantly contributes to the reduction of greenhouse gas generation. With its high calorific value, this biochar has the potential to lessen reliance on fossil fuels for energy production. All things considered, biomass gasification technology is essential for lowering greenhouse gas emissions and advancing the production of sustainable energy.

6.5 Stages of pyrolysis and gasification

The various stages of pyrolysis and gasification are depicted in Fig. 6.3.

6.5.1 Drying

Drying is a kind of pretreatment that lowers the moisture content of the feedstock to lessen agglomeration and fluidization problems and increase conversion efficiency. At temperatures between 100°C and 200°C during the drying stage, the free and restricted moisture in the feedstock is

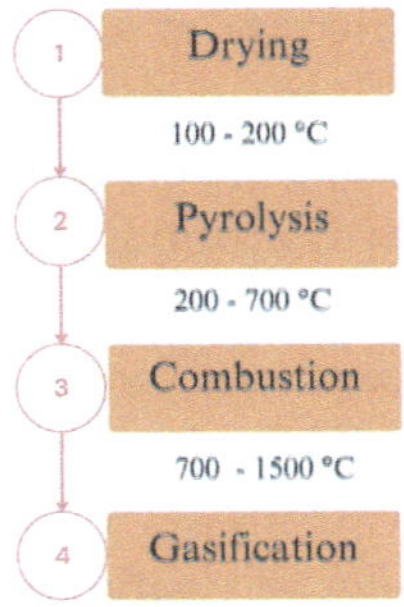

Fig. 6.3 The stages of pyrolysis and gasification.

dehydrated; this is insufficient to cause the feedstock to thermally disintegrate (Zhang et al., 2022; Sharma et al., 2022a, 2022b).

6.5.2 Pyrolysis and combustion

The dried feedstock is broken down into charcoal and other volatiles during the pyrolysis stage at a high temperature (200°C to 700°C) and without oxygen. At the combustion stage, which lasts between 700°C and 1500°C, the ingredients are completely or partially oxidized into H_2O, CO_2, and CO. When biochar combines with water, carbon dioxide, and oxygen, it creates CO, H_2, CH_4, acetylene, and ethylene (Zhang et al., 2022; Sharma et al., 2022a, 2022b).

6.5.3 Gasification

For gasification, notable gasifying substances include steam, air, oxygen, carbon dioxide, and their mixes. Despite the fact that air is commonly used due to its affordability and accessibility. The physicochemical properties, product composition, heating value, and biomass conversion efficiency of biochar are all enhanced by the introduction of gasifying agents. Gasification is a beneficial technique that can be used for resource recovery, resource conversion, biofuel production, biochar, biohydrogen, syngas, biomethane, and other useful substances (Ajorloo et al., 2022; Dafiqurrohman et al., 2022; Sharma et al., 2022a, 2022b).

6.6 Scalability/commercialization of pyrolysis and gasification

The bio-oil production process is constrained, yet it offers a possible replacement for fuel derived from petroleum. Some examples of novel reactor designs for biomass pyrolysis include FBs, revolving cones, fixed beds, vacuum pyrolyzers, screw feeders, heated kilns, ablative, etc. Three reactor designs, including revolving cones and FBs that bubble and circulate, have been industrialized. Other approaches, such as integrated pyrolysis processes including iron ore reduction and NO_x reduction techniques, hydrothermal pyrolysis, catalytic pyrolysis, and microwave-assisted pyrolysis, are also being studied for commercialization (Nagendranatha Reddy and Mohan, 2016; Reddy et al., 2016; Khosravanipour Mostafazadeh et al., 2018). The laboratory trials must be expanded to take into account issues including heat integration technology, reactor type, and processing mode. Further research is needed to understand the impacts of altered operating parameters and optimize the catalytic FP's heat balance. It is also crucial to remove solid particles from the final produce, oil, to lessen the adverse impacts (Yildiz et al., 2016).

In an effort to construct a biomass rapid pyrolysis unit on an industrial scale, a downdraft circulating FB reactor was utilized. The plant was operated constantly without the use of external energy and has a feeding system, a heat carrier system, a reactor system, a cyclone system, a condensation system, and a system for separating carbon. The bio-oil output reached 48.1 wt% at 550°C, with phenolic compounds constituting the majority of the oil. For char and noncondensable products, the plant's efficiency was kept at 26.0 and 25.9 wt%, respectively (Cai and Liu, 2016).

Similar to pyrolysis, large-scale gasification entails a number of procedures and yields flammable gas. In contrast to pyrolysis, gasification is employed in the same facilities rather than being stored or transported. By substituting fossil fuels and eliminating extraction, it lowers CO_2 emissions. In addition to using materials now disposed of in landfills, gasification offers greater electrical efficiency than direct trash combustion. The advantages of waste gasification are numerous (Chavando et al., 2021).

In conclusion, research into novel reactor designs and pyrolysis techniques is being done in order to commercialize the bio-oil production process, which offers a competitive substitute for fuels obtained from petroleum. Increasingly, studies are being conducted to improve the heat balance in catalytic FP, focusing on topics including heat integration, reactor design, and processing modes. To lessen negative effects, it's also essential to remove particulates from the product oil. Moreover, the accomplishment of noteworthy bio-oil yields through the effective application of a downdraft circulating FB reactor proved the viability of rapid pyrolysis of biomass on a commercial scale (Tables 6.1 and 6.2).

Table 6.1 Types of Thermochemical processes and their methodology.

Thermochemical Methods	Agro-waste	Temperature (°C)	Heating rate (°C/min)	Residence time (min)	O_2 requirement	Other requirements	References
SP	Brown peat and coir	300-600	5–7	60–120	No	–	Méndez et al. (2015)
FP	Sawdust	500	300	0–20	No	–	Laird et al. (2017)
FLP	Maize-cob biomass waste	100	1000	0–1	No	–	Adelawon et al. (2022)
VP	Maize-cob biomass waste	450-600	low	–	No	0.05–0.2 MPa Pressure	Adelawon et al. (2022)
Gasification	Olive oil waste	700–900	–	–	–	Catalysts $ZnCl_2$	González et al. (2008)

Table 6.2 Advantages and disadvantages of various thermochemical processes.

Thermochemical method	Advantage	Disadvantage	References
SP	Moderate temperature and minimal energy usage	Low product yield, lengthy residence period, and slower process	Méndez et al. (2015)
FP	Better product recovery, a high conversion rate, and a short residency duration	Low biochar yield	Laird et al. (2017)
FLP	Quick, effective and efficient procedure	An excessively high reaction temperature and a very slow heating rate	Adelawon et al. (2022)
VP	Superior product quality and increased yield	High expense of keeping a vacuum environment	Adelawon et al. (2022)
Gasification	Producing acetylene, ethylene, and other beneficial compounds helps to maintain the environment and produces fuels that are eco-friendly and renewable.	High cost of energy	You et al. (2018)

Furthermore, large-scale gasification, which produces flammable gas, provides a sustainable and green method of disposing of trash while lowering CO_2 emissions. Gasification makes use of materials that are now disposed of in landfills, which helps to make waste management more effective and environmentally friendly. These developments in gasification and pyrolysis technology portend a more environmentally friendly and sustainable energy future (Mohan et al., 2016).

6.7 Conclusion and future prospects

Gasification and pyrolysis of agricultural residues have extremely bright future potential. The use of these technologies holds great promise for contributing significantly to the shift towards a more sustainable and circular economy. The need for renewable energy is expanding quickly on a worldwide scale as nations look to lessen their dependency on fossil fuels and slow down climate change. This increasing demand can be met in part by producing biofuels, heat, and power through the pyrolysis and gasification of agricultural leftovers. Burning is one of the traditional techniques of managing agricultural residues, which can have a harmful influence on the environment. An environmentally acceptable and more sustainable option for managing agricultural wastes is to use gasification and pyrolysis.

To sum up, it is critical to investigate biomass pyrolysis and gasification as renewable energy sources. Many variables that affect the results of these processes have been covered in this book chapter, including temperature, pyrolysis conditions, gasification conditions, biomass composition, and particle size. It has also highlighted the potential applications of catalysts and parameter modification for gaseous product production. This investigation provides a clear picture of the biomass pyrolysis and gasification technologies' economic feasibility and adaptability. These methods provide a variety of opportunities for utilizing various biomass wastes to produce a range of valuable end products; they are not limited to any particular feedstocks or end products. The enormous potential of biomass as a sustainable energy source is highlighted by this article, opening the door to a more resourceful and environmentally friendly future.

Acknowledgments

The authors thank the Principal and Management of CBIT for their constant support and encouragement in carrying out this work.

References

Adelawon, B.O., Latinwo, G.K., Eboibi, B.E., Agbede, O.O., Agarry, S.E., 2022. Comparison of the slow, fast, and flash pyrolysis of recycled maize-cob biomass waste, box-benhken process optimization and characterization studies for the thermal fast pyrolysis production of bio-energy. Chem. Eng. Commun. 209 (9), 1246–1276.

Ahmad, M., Rajapaksha, A.U., Lim, J.E., Zhang, M., Bolan, N., Mohan, D., Vithanage, M., Lee, S.S., Ok, Y.S., 2014. Biochar as a sorbent for contaminant management in soil and water: a review. Chemosphere. 1879-1298. 99, 19–33. Available from: https://www.elsevier.com/locate/chemosphere. doi:10.1016/j.chemosphere.2013.10.071.

Ahmed, G., Acharya, S., Kawale, H., Singh, A., Kishore, N., Pal, S., 2020. Thermochemical conversion of *Polyalthia longifolia* leaves at different temperatures and characterization of their products. Fuel 280, 118574. doi:10.1016/j.fuel.2020.118574.

Ajorloo, M., Ghodrat, M., Scott, J., Strezov, V., 2022. Recent advances in thermodynamic analysis of biomass gasification: a review on numerical modelling and simulation. J. Energy Inst. 1743-9671. 102, 395–419. doi:10.1016/j.joei.2022.05.003.

Akhtar, J., Amin, N.S., 2012. A review on operating parameters for optimum liquid oil yield in biomass pyrolysis. Renew. Sustain. Energy Rev. 16 (7), 5101–5109.

Anca-Couce, A., 2016. Reaction mechanisms and multi-scale modelling of lignocellulosic biomass pyrolysis. Prog. Energy Combust. Sci. 53, 41–79.

Andrade, L.A., Batista, F.R.X., Lira, T.S., Barrozo, M.A.S., Vieira, L.G.M., 2018. Characterization and product formation during the catalytic and non-catalytic pyrolysis of the green microalgae *Chlamydomonas reinhardtii*. Renew. Energy 119, 731–740.

Anukam, A., Berghel, J., 2021. Biomass pretreatment and characterization: a review. In: Biotechnological Applications of Biomass. IntechOpen. doi:10.5772/intechopen.93607.

Asomaning, J., Haupt, S., Chae, M., Bressler, D.C., 2018. Recent developments in microwave-assisted thermal conversion of biomass for fuels and chemicals. Renew. Sustain. Energy Rev. 92, 642–657.

Athira, G., Bahurudeen, A., Appari, S., 2019. Sustainable alternatives to carbon intensive paddy field burning in India: a framework for cleaner production in agriculture, energy, and construction industries. J. Clean. Prod. 236, 117598.

Bamido, A.O., 2018. Design of a fluidized bed reactor for biomass pyrolysis (Doctoral dissertation, University of Cincinnati).

Barati, Z., Latif, S., Romuli, S., Müller, J., 2020. Enzyme-assisted mechanical peeling of Cassava Tubers. Catalysts 2073-4344, 10 (1), 66. doi:10.3390/catal10010066.

Baruah, D., Baruah, D., 2014. Modeling of biomass gasification: a review. Renew. Sustain. Energy Rev. 39, 806–815. doi:10.1016/j.rser.2014.07.129.

Beohar, H., Gupta, B., Sethi, V.K., Pandey, M., Parmar, H., 2012. Effect of air velocity, fuel rate and moisture content on the performance of updraft biomass gasifier using fluent tool. Int. J. Mod. Eng. Res. 2 (5), 3622–3627.

Bhaskar, T., Ruan, R., Park, Y.K., Yang, H., Chen, G., 2020. Pyrolysis, combustion and gasification of biomass (PCGB-2020). Bioresour. Technol. 313, 123803. doi:10.1016/j.biortech.2020.123803.

Bhoi, P.R., Ouedraogo, A.S., Soloiu, V., Quirino, R., 2020. Recent advances on catalysts for improving hydrocarbon compounds in bio-oil of biomass catalytic pyrolysis. Renew. Sustain. Energy Rev. 1364-0321. 121, 109676. doi:10.1016/j.rser.2019.109676.

Bhuvaneshwari, S., Hettiarachchi, H., Meegoda, J., 2019. Crop residue burning in india: policy challenges and potential solutions. Int. J. Environ. Res. Public Health 16 (5), 832. doi:10.3390/ijerph16050832.

Biswas, B., Pandey, N., Bisht, Y., Singh, R., Kumar, J., Bhaskar, T., 2017. Pyrolysis of agricultural biomass residues: comparative study of corn cob, wheat straw, rice straw and rice husk. Bioresour. Technol. 0960-8524. 237, 57–63. doi:10.1016/j.biortech.2017.02.046.

Cai, W., Liu, R., 2016. Performance of a commercial-scale biomass fast pyrolysis plant for bio-oil production. Fuel. 0016-2361. 182, 677–686. doi:10.1016/j.fuel.2016.06.030.

Campuzano, F., Ortíz, C.F., Betancur, M., Martínez, J.D., 2018. Characterization of three different solid wastes as energy resources for pyrolysis. IOP Conf. Ser. Mater. Sci. Eng. 437 (1), 012002.

Ceranic, M., Kosanic, T., Djuranovic, D., Kaludjerovic, Z., Djuric, S., Gojkovic, P., Bozickovic, R., 2016. Experimental investigation of corn cob pyrolysis. J. Renew. Sustain. Energy. 1941-7012. 8 (6). doi:10.1063/1.4966695.

Chakraborty, I., Sathe, S.M., Dubey, B.K., Ghangrekar, M.M., 2020. Waste-derived biochar: applications and future perspective in microbial fuel cells. Bioresour. Technol. 0960-8524. 312, 123587. doi:10.1016/j.biortech.2020.123587.

Chavando, J.A.M., Silva, V., Guerra, D.R.D.S., Eusébio, D., Cardoso, J.S., Tarelho, L.A., 2021. Review chapter: waste to energy through pyrolysis and gasification in brazil and mexico. Gasification 3. doi:10.5772/intechopen.98383.

Chen, J., Wang, Y., Lang, X., Fan, S., 2017. Evaluation of agricultural residues pyrolysis under non-isothermal conditions: thermal behaviors, kinetics, and thermodynamics. Bioresour. Technol. 241, 340–348.

Choe, K., 2021. Review of wood biomass cyclone burner. Energies 14 (16), 4807. doi:10.3390/en14164807.

Colantoni, A., Evic, N., Lord, R., Retschitzegger, S., Proto, A.R., Gallucci, F., Monarca, D., 2016. Characterization of biochars produced from pyrolysis of pelletized agricultural residues. Renew. Sustain. Energy Rev. 64, 187–194.

Dafiqurrohman, H., Safitri, K.A., Setyawan, M.I.B., Surjosatyo, A., Aziz, M., 2022. Gasification of rice wastes toward green and sustainable energy production: a review. J. Clean. Prod.. 0959-6526. 366. Available from: https://www.journals.elsevier.com/journal-of-cleaner-production. doi:10.1016/j.jclepro.2022.132926.

De Pretto, C., Giordano, R.D.L.C., Tardioli, P.W., Costa, C.B.B., 2018. Possibilities for producing energy, fuels, and chemicals from soybean: a biorefinery concept. Waste Biomass Valoriz. 1877-265X. 9 (10), 1703–1730. Available from: https://www.springer.com/engineering/journal/12649. doi:10.1007/s12649-017-9956-3.

Edison, S., Anantharaman, M., Srinivas, T., 2006. Introduction: Status of cassava in India, an overall view. Central Tuber Crops Research Institute, Indian Council of Agricultural Research, Thiruvananthapuram, India. pp. 11–16.

Edrich, R., Bradley, T., Graboski, M.S., 1985. The gasification of ponderosa pine charcoal. In:Overend, R.P., Milne, T.A., Mudge, L.K. (Eds.), Fundamentals of Thermochemical Biomass Conversion. Springer, Dordrecht, pp. 557–566. doi:10.1007/978-94-009-4932-4_31.

Eke, J., Onwudili, J.A., Bridgwater, A.V., 2020. Influence of moisture contents on the fast pyrolysis of trommel fines in a bubbling fluidized bed reactor. Waste Biomass Valoriz. 11, 3711–3722.

Gabhane, J.W., Bhange, V.P., Patil, P.D., Bankar, S.T., Kumar, S., 2020. Recent trends in biochar production methods and its application as a soil health conditioner: a review. SN Appl. Sci.. 2523-3971. 2 (7). Available from: https://www.springer.com/snas. doi:10.1007/s42452-020-3121-5.

Ginni, G., Kavitha, S., Kannah, R.Y., Bhatia, S.K., Kumar, S.A., Rajkumar, M., Kumar, G., Pugazhendhi, A., Chi, N.T.L., Banu, J.R., 2021. Valorization of agricultural residues: different biorefinery routes. J. Environ. Chem. Eng. 2213-3437. 9 (4). Available from: https://www.journals.elsevier.com/journal-of-environmental-chemical-engineering/. doi:10.1016/j.jece.2021.105435.

Glushkov, D., Nyashina, G., Shvets, A., Pereira, A., Ramanathan, A., 2021. Current status of the pyrolysis and gasification mechanism of biomass. Energies. 1996-1073. 14 (22), 7541. doi:10.3390/en14227541.

González, J.F., Román, S., Bragado, D., Calderón, M., 2008. Investigation on the reactions influencing biomass air and air/steam gasification for hydrogen production. Fuel Process. Technol. 89 (8), 764–772.

Guo, H., Zhang, S., Kou, Z., Zhai, S., Ma, W., Yang, Y., 2015. Removal of cadmium(II) from aqueous solutions by chemically modified maize straw. Carbohydr. Polym. 0144-8617. 115, 177–185. doi:10.1016/j.carbpol.2014.08.041.

Guzmán, A., Delvasto, A.S., Sánchez, V.E., 2015. Valorization of rice straw waste: an alternative ceramic raw material. Cerâmica 61 (357), 126–136. doi:10.1590/0366-69132015613571888.

Hagner, M., Tiilikkala, K., Lindqvist, I., Niemelä, K., Wikberg, H., Källi, A., Rasa, K., 2020. Performance of liquids from slow pyrolysis and hydrothermal carbonization in plant protection. Waste Biomass Valoriz. 11 (3), 1005–1016.

Haris, M., Hamid, Y., Usman, M., Wang, L., Saleem, A., Su, F., Guo, J., Li, Y., 2021. Crop-residues derived biochar: synthesis, properties, characterization and application for the removal of trace elements in soils. J. Hazard. Mater. 0304-3894. 416, 126212. doi:10.1016/j.jhazmat.2021.126212.

Havilah, P.R., Sharma, A.K., Govindasamy, G., Matsakas, L., Patel, A., 2022. Biomass gasification in downdraft gasifiers: a technical review on production, up-gradation and application of synthesis gas. Energies. 1996-1073. 15 (11). Available from: https://www.mdpi.com/1996-1073/15/11/3938/pdf?version=1653637886. doi:10.3390/en15113938.

Himmel, M.E., Ding, S.Y., Johnson, D.K., Adney, W.S., Nimlos, M.R., Brady, J.W., Foust, T.D., 2007. Biomass recalcitrance: engineering plants and enzymes for biofuels production. Science 315 (5813), 804–807.

Hong, Z., Zhong, F., Niu, W., Zhang, K., Su, J., Liu, J., Wu, F., 2020. Effects of temperature and particle size on the compositions, energy conversions and structural characteristics of pyrolysis products from different crop residues. Energy 190, 116413.

Ighalo, J.O., Iwuchukwu, F.U., Eyankware, O.E., Iwuozor, K.O., Olotu, K., Bright, O.C., Igwegbe, C.A., 2022. Flash pyrolysis of biomass: a review of recent advances. Clean Technol. Environ. Policy. 1618-9558. Available from: https://www.link.springer.com/journal/10098. doi:10.1007/s10098-022-02339-5.

Jugwanth, Y., Sewsynker-Sukai, Y., Kana, E G., 2020. Valorization of sugarcane bagasse for bioethanol production through simultaneous saccharification and fermentation: optimization and kinetic studies. Fuel 262, 116552.

Khan, T.S., Mubeen, U., 2012. Wheat straw: a pragmatic overview. Curr. Res. J. Biol. Sci 4 (6), 673–675.

Khosravanipour Mostafazadeh, A., Solomatnikova, O., Drogui, P., Tyagi, R.D., 2018. A review of recent research and developments in fast pyrolysis and bio-oil upgrading. Biomass Convers. Biorefin. 8, 739–773.

Kong, L., Zhang, L., Junlin, G., Gou, L., Xie, L., Wang, Y., Dai, L., 2020b. Catalytic hydrotreatment of kraft lignin into aromatic alcohols over nickel-rhenium supported on niobium oxide catalyst. Bioresour. Technol. 0960-8524. 299, 122582. doi:10.1016/j.biortech.2019.122582.

Kremling, M., Briesemeister, L., Gaderer, M., Fendt, S., 2017. Hartmut spliethoff, oxygen-blown entrained flow gasification of biomass: impact of fuel parameters and oxygen stoichiometric ratio. Energy Fuels. 0887-0624. 31 (4), 3949–3959. doi:10.1021/acs.energyfuels.6b02949.

Laird, D.A., Novak, J.M., Collins, H.P., Ippolito, J.A., Karlen, D.L., Lentz, R.D., Sistani, K.R., Spokas, K., Van Pelt, R.S., 2017. Multi-year and multi-location soil quality and crop biomass yield responses to hardwood fast pyrolysis biochar. Geoderma. 0016-7061. 289, 46–53. doi:10.1016/j.geoderma.2016.11.025.

Li, H., Wang, Y., Zhou, N., Dai, L., Deng, W., Liu, C., Cheng, Y., Liu, Y., Cobb, K., Chen, P., Ruan, R., 2021. Applications of calcium oxide–based catalysts in biomass pyrolysis/gasification—a review. J. Clean. Prod. 291, 125826. doi:10.1016/j.jclepro.2021.125826.

Li, Y., Gupta, R., Zhang, Q., You, S., 2022. Review of biochar production via crop residue pyrolysis: development and perspectives. Bioresour. Technol. 369, 128423.

Liu, Z., Niu, W., Chu, H., Niu, Z., 2018. Process optimization for straws pyrolysis and analysis of biochar physiochemical properties. Trans. Chin. Soc. Agric. Eng. 34 (5), 196–203.

Luo, D., Wang, L., Nan, H., Cao, Y., Wang, H., Kumar, T.V., Wang, C., 2023. Phosphorus adsorption by functionalized biochar: a review. Environ. Chem. Lett. 1610-3661. 21 (1), 497–524. Available from: https://www.springer.com/journal/10311. doi:10.1007/s10311-022-01519-5.

Makavana, J.M., Kalaiya, S.V., Dulawat, M.S., Sarsavadiya, P.N., Chauhan, P.M., 2020. Development and performance evaluation of batch type biomass pyrolyser for agricultural residue. Biomass Convers. Biorefin. 12 (11), 5067–5074. doi:10.1007/s13399-020-01105-1.

Méndez, A., Paz-Ferreiro, J., Gil, E., Gascó, G., 2015. The effect of paper sludge and biochar addition on brown peat and coir based growing media properties. Sci. Hortic. 0304-4238. 193, 225–230. doi:10.1016/j.scienta.2015.07.032.

Milne, T., Agblevor, F., Davis, M., Deutch, S., Johnson, D., 1997. A review of the chemical composition of fast-pyrolysis oils from biomass. In: Developments in Thermochemical Biomass Conversion. Springer Science and Business Media LLC, pp. 409–424. doi:10.1007/978-94-009-1559-6_32.

Mishra, B., Mohanta, Y.K., Nagendranatha Reddy, C., Reddy, S.D.M., Mandal, S.K., Yadavally, R., Sarma, H., 2023. Valorization of agro-industrial biowaste to biomaterials: an innovative circular bioeconomy approach. Circular Econ. 2, 100050. doi:10.1016/j.cec.2023.100050.

Mohan, D., Sarswat, A., Ok, Y.S., Pittman Jr, C.U., 2014. Organic and inorganic contaminants removal from water with biochar, a renewable, low cost and sustainable adsorbent–a critical review. Bioresour. Technol. 160, 191–202.

Mohan, S.V., Nikhil, G.N., Chiranjeevi, P., Reddy, C.N., Rohit, M.V., Kumar, A.N., Sarkar, O., 2016. Waste biorefinery models towards sustainable circular bioeconomy: critical review and future perspectives. Bioresour. Technol. 1873-2976. 215, 2–12. Available from: https://www.elsevier.com/locate/biortech. doi:10.1016/j.biortech.2016.03.130.

Nagendranatha Reddy, C., Mehariya, S., Kavitha, R.Y.K., Jayaprakash, K., Yadavalli, R., Banu, R., Karthikeyan, O.P., 2020. Electro-fermentation of biomass for high-value organic acids. In: Ed., Verma, P.. Biorefineries: A Step Towards Renewable and Clean Energy. Clean Energy Production Technologies, Springer, Singapore, pp. 417–436. doi:10.1007/978-981-15-9593-6_16.

Nagendranatha Reddy, C., Min, B., 2019. Biological conversion of food waste to value addition in Microbial Fuel Cell. In: Singh, L., et al., (Ed.), Waste to Sustainable Energy: MFCs-Prospects through Prognosis. CRC Press, pp. 212–239. doi:10.1201/9780429448799. 9781138328211.

Nagendranatha Reddy, C., Modestra, J.A., Kumar, A.N., Mohan, S.V., 2016. Waste Remediation Integrating with Value Addition: biorefinery Approach Towards Sustainable Bio-based Technologies. Springer India.231-256. In: Kalia, V.C. (Ed.), Microbial Factories, Biofuels, Waste Treatment: Volume 1. doi:10.1007/978-81-322-2598-0_14.

Nagendranatha Reddy, C., Mohan, S.V., 2016. Integrated bio-electrogenic process for bio-electricity production and cathodic nutrient recovery from azo dye wastewater. Renew. Energy 98, 188–196.

Nagendranatha Reddy, C., Vineetha, Y., Priyanka, A., Shalini, A., B.Mishra, Y.R., Swapna, V., 2022. Techno-economic and environmental impact analysis of biofuels produced from microalgal biomass. Biofuels Bioenergy. doi:10.1016/B978-0-323-90040-9.00006-0.

Narnaware, S.L., Vahora, S., 2022. Study on physico-chemical treatment of effluents from biomass gasifier power generation system. Int. J. Environ. Sustain. Dev. 21 (1-2), 138–149.

Naryanto, R.F., Enomoto, H., Cong, A.V., Fukadu, K., Zong, Z., Delimayanti, M.K., Chunti, C., Noda, R., 2020. The effect of moisture content on the tar characteristic of wood pellet feedstock in a downdraft gasifier. Appl. Sci. 2076-3417. 10 (8), 2760. doi:10.3390/app10082760.

Oh, S., Lee, J., Lam, S.S., Kwon, E.E., Ha, J.M., Tsang, D.C.W., Ok, Y.S., Chen, W.H., Park, Y.K., 2021. Fast hydropyrolysis of biomass conversion: a comparative review. Bioresour. Technol.. 1873-2976. 342. Available from: https://www.elsevier.com/locate/biortech. doi:10.1016/j.biortech.2021.126067.

Osman, A.I., Mehta, N., Elgarahy, A.M., Al-Hinai, A., Al-Muhtaseb, A.H., Rooney, D.W., 2021. Conversion of biomass to biofuels and life cycle assessment: a review. Environ.

Chem. Lett. 1610-3661. 19 (6), 4075–4118. Available from: http://springerlink.metapress.com/app/home/journal.asp?wasp=d86tgdwvtg0yvw9gvkwp&referrer=parent&backto=browsepublicationsresults. doi:10.1007/s10311-021-01273-0.

Patra, B.R., Nanda, S., Dalai, A.K., Meda, V., 2021. Slow pyrolysis of agro-food wastes and physicochemical characterization of biofuel products. Chemosphere. 1879-1298. 285. Available from: https://www.elsevier.com/locate/chemosphere. doi:10.1016/j.chemosphere.2021.131431.

Pecha, M.B., Garcia-Perez, M., 2020. Pyrolysis of lignocellulosic biomass: oil, char, and gas. In: Bioenergy: Biomass to Biofuels and Waste to Energy. Elsevier, United States, pp. 581–619. Available from: https://www.elsevier.com/books/bioenergy/dahiya/978-0-12-815497-7. doi:10.1016/B978-0-12-815497-7.00029-4.

Perea-Moreno, M.-A., Manzano-Agugliaro, F., Hernandez-Escobedo, Q., Perea-Moreno, A.-J., 2018. Peanut shell for energy: properties and its potential to respect the environment. Sustainability. 2071-1050. 10 (9), 3254. doi:10.3390/su10093254.

Qu, T., Guo, W., Shen, L., Xiao, J., Zhao, K., 2011. Experimental study of biomass pyrolysis based on three major components: hemicellulose, cellulose, and lignin. Ind. Eng. Chem. Res. 50 (18), 10424–10433.

Rahimi-Ajdadi, F., Asli-Ardeh, E., Ahmadi-Ara, A., 2018. Effect of varying parboiling conditions on head rice yield for common paddy varieties in Iran. Acta Technol. Agric. 21 (1), 1–7. doi:10.2478/ata-2018-0001.

Rahimi, Z., Anand, A., Gautam, S., 2022. An overview on thermochemical conversion and potential evaluation of biofuels derived from agricultural wastes. Energy Nexus 7, 100125.

Ramanan, M.V., Lakshmanan, E., Sethumadhavan, R., Renganarayanan, S., 2008. Modeling and experimental validation of cashew nut shell char gasification adopting chemical equilibrium approach. Energy Fuels 22 (3), 2070–2078.

Ranum, P., Peña-Rosas, J.P., Garcia-Casal, M.N., 2014. Global maize production, utilization, and consumption. Ann. N.Y. Acad. Sci. 1312 (1), 105–112.

Reddy, C.N., Modestra, J.A., Kumar, A.N., Mohan, S.V., 2016. Waste remediation integrating with value addition: biorefinery approach towards sustainable bio-based technologies. In: Microbial Factories: Biofuels, Waste Treatment: Volume 1. Springer, India, pp. 231–256. doi:10.1007/978-81-322-2598-0_14.

Reddy, S.D.M., Deepika, N., Dropathi, M.R., Vishwanutha, S., Daaman, J.D., Nagendranatha Reddy, C., Yadavalli, R., 2023. Factors affecting the microalgal biomass productivity in photobioreactors. In: Sirohi, R., Pandey, A., Sim, S.J., Chang, J.-S., Lee, D.-J. (Eds.), Current Developments in Biotechnology and Bioengineering: Photobioreactors: Design and Applications. Elsevier, ISBN: 978-0-323-99911-3. Available from: https://www.sciencedirect.com/science/article/pii/B9780323999113000038.

Reza, M.S., Azad, A.K., Bakar, M.S.A., Karim, M.R., Sharifpur, M., Taweekun, J., 2022. Evaluation of thermochemical characteristics and pyrolysis of fish processing waste for renewable energy feedstock. Sustainability. 2071-1050. 14 (3), 1203. doi:10.3390/su14031203.

Reza, M.S., Baktybaevna, I.Z., Afroze, S., Kuterbekov, K., Kabyshev, A., Bekmyrza, K.Z., Islam, M.S., 2023a. Influence of catalyst on the yield and quality of bio-oil for the catalytic pyrolysis of biomass: a comprehensive review. Energies 16 (14), 5547.

Reza, M.S., Taweekun, J., Afroze, S., Siddique, S.A., Islam, M.S., Wang, C., Azad, A.K., 2023b. Investigation of thermochemical properties and pyrolysis of barley waste as a source for renewable energy. Sustainability 15 (2), 1643.

Röder, M., Mohr, A., Liu, Y., 2020. Sustainable bioenergy solutions to enable development in low- and middle-income countries beyond technology and energy access. Biomass Bioenergy. doi:10.1016/j.biombioe.2020.105876.

Rong, L., Maneerung, T., Ng, J.C., Neoh, K.G., Bay, B.H., Tong, Y.W., Wang, C.H., 2015. Co-gasification of sewage sludge and woody biomass in a fixed-bed downdraft gasifier: toxicity assessment of solid residues. Waste Manag. 36, 241–255.

Roy, P., Dias, G., 2017. Prospects for pyrolysis technologies in the bioenergy sector: a review. Renew. Sustain. Energy Rev. 1364-0321. 77, 59–69. doi:10.1016/j.rser.2017.03.136.

Sahoo, K., Kumar, A., Chakraborty, J.P., 2021. A comparative study on valuable products: bio-oil, biochar, non-condensable gases from pyrolysis of agricultural residues. J. Mater. Cycles Waste Manag. 23, 186–204.

Saifuddin, N., Salman, B., Refal, H., Ong, M., 2017. Microwave pyrolysis of lignocellulosic biomass-a contribution to power Africa. Energy, Sustain. Soc. 7 (23), 1–24. doi:10.1186/s13705-017-0126-z.

Saikia, R., Baruah, B., Kalita, D., Pant, K.K., Gogoi, N., Kataki, R., 2018. Pyrolysis and kinetic analyses of a perennial grass (*Saccharum ravannae* L.) from north-east India: optimization through response surface methodology and product characterization. Bioresour. Technol. 1873-2976. 253, 304–314. Available from: https://www.elsevier.com/locate/biortech. doi:10.1016/j.biortech.2018.01.054.

Sekoai, P.T., Daramola, M.O., 2015. Biohydrogen production as a potential energy fuel in South Africa. Biofuel Res. J. 22928782. 2 (2), 223–226. doi:10.18331/BRJ2015.2.2.3.

Shahryari, Z., Fazaelipoor, M.H., Setoodeh, P., Nair, R.B., Taherzadeh, M.J., Ghasemi, Y., 2018. Utilization of wheat straw for fungal phytase production. Int. J. Recycl. Org. Waste Agric. 2251-7715. 7 (4), 345–355. Available from: https://www.springer.com/environment/pollution+and+remediation/journal/40093. doi:10.1007/s40093-018-0220-z.

Sharma, A., Pareek, V., Zhang, D., 2015. Biomass pyrolysis—A review of modelling, process parameters and catalytic studies. Renew. Sustain. Energy Rev. 50, 1081–1096.

Sharma, P., Sen, S., Sheth, P.N., Mohapatra, B.N., 2022a. Multizone model of a refused derived fuel gasification: a thermodynamic semi-empirical approach. Energy Convers. Manag. 260, 115621.

Sharma, S., Rana, V.S., Rana, N., Prasad, H., Sharma, U., Patiyal, V., 2022b. Biochar from fruit crops waste and its potential impact on fruit crops. Sci. Hortic. 299, 111052. doi:10.1016/j.scienta.2022.111052.

Sharma, S., Sheth, P.N., 2016. Air–steam biomass gasification: experiments, modeling and simulation. Energy Convers. Manag. 110, 307–318.

Sidana, A., Farooq, U., 2014. Sugarcane Bagasse: a potential medium for fungal cultures. Chinese J. Biol. 2314-7474. 2014, 1–5. doi:10.1155/2014/840505.

Sikarwar, V.S., Zhao, M., Clough, P., Yao, J., Zhong, X., Memon, M.Z., Shah, N., Anthony, E.J., Fennell, P.S., 2016. An overview of advances in biomass gasification. Energy Environ. Sci. 1754-5706. 9 (10), 2939–2977. Available from: https://www.rsc.org/Publishing/Journals/EE/About.asp. doi:10.1039/c6ee00935b.

Solís Hontañón, S., 2016. Biochar production by biomass pyrolysis for various applications.

Suhartono, B.D.P., Azizah, I.N., 2006. Synthetic gas (syngas) production in downdraft corncob gasifier and its application as fuel using conventional domestic (LPG) stove. Combustion 5, 7.

Tagade, A., Kirti, N., Sawarkar, A.N., 2021. Pyrolysis of agricultural crop residues: an overview of researches by Indian scientific community. Bioresour. Technol. Rep. 15, 100761.

Tripathi, M., Sahu, J.N., Ganesan, P., 2016. Effect of process parameters on production of biochar from biomass waste through pyrolysis: a review. Renew. Sustain. Energy Rev. 55, 467–481.

Tumuluru, J.S., Ed.. 2017. Biomass Volume Estimation and Valorization for Energy. BoD–Books on Demand.

Uddin, M.N., Techato, K., Taweekun, J., Rahman, M.M., Rasul, M.G., Mahlia, T.M.I., Ashrafur, S.M., 2018. An overview of recent developments in biomass pyrolysis technologies. Energies. 1996-1073. 11 (11). Available from: https://www.mdpi.com/1996-1073/11/11. doi:10.3390/en11113115.

UN, 2022. World population projected to reach 9.8 billion in 2050, and 11.2 billion in 2100. Department of Economic and Social Affair, United Nations, https://www.un.org/en/desa/world-population-projected-reach-98-billion-2050-and-112-billion-2100 (Accessed 12 January, 2022).

Varma, A.K., Mondal, P., 2017. Pyrolysis of sugarcane bagasse in semi batch reactor: effects of process parameters on product yields and characterization of products. Ind. Crops Prod. 95, 704–717.

Venkata Mohan, S., Nikhil, G.N., Chiranjeevi, P., Reddy, C.N., Rohit, M.V., Kumar, A.N., Sarkar, O., 2016. Waste biorefinery models towards sustainable bioeconomy: critical review and future perspectives. Bioresour. Technol. 215, 2–12.

Vivek, N., Nair, L.M., Mohan, B., Nair, S.C., Sindhu, R., Pandey, A., Shurpali, N., Binod, P., 2019. Bio-butanol production from rice straw—Recent trends, possibilities, and challenges. Bioresour. Technol. Rep. 2589-014X (7). Available from: https://www.journals.elsevier.com/bioresource-technology-reports. doi:10.1016/j.biteb.2019.100224.

Wang, A., Song, H., 2018. Maximizing the production of aromatic hydrocarbons from lignin conversion by coupling methane activation. Bioresour. Technol. 1873-2976. 268, 505–513. Available from: https://www.elsevier.com/locate/biortech. doi:10.1016/j.biortech.2018.08.026.

Wang, S., Chen, J., Zhang, F., Wang, Y., 2017. Chapter 6. Characterization and separation of bio-oil. In: Green Chemistry Series. Royal Society of Chemistry (RSC), pp. 96–116. doi:10.1039/9781788010245-00096.

Woods, J., Williams, A., Hughes, J.K., Black, M., Murphy, R., 2010. Energy and the food system. Philos. Trans. R. Soc. B Biol. Sci. 1471-2970. 365 (1554), 2991–3006. Available from: https://rstb.royalsocietypublishing.org/content/365/1554/2991.full.pdf+html. doi:10.1098/rstb.2010.0172.

Wu, C.Z., Yin, X.L., Ma, L.L., Zhou, Z.Q., Chen, H.P., 2009. Operational characteristics of a 1.2-MW biomass gasification and power generation plant. Biotechnol. Adv. 0734-9750. 27 (5), 588–592. doi:10.1016/j.biotechadv.2009.04.020.

Yildiz, G., Ronsse, F., Duren, R.V., Prins, W., 2016. Challenges in the design and operation of processes for catalytic fast pyrolysis of woody biomass. Renew. Sustain. Energy Rev. 1879-0690. 57, 1596–1610. Available from: https://www.journals.elsevier.com/renewable-and-sustainable-energy-reviews. doi:10.1016/j.rser.2015.12.202.

You, S., Ok, Y.S., Tsang, D.C.W., Kwon, E.E., Wang, C.H., 2018. Towards practical application of gasification: a critical review from syngas and biochar perspectives. Crit. Rev. Environ. Sci. Technol. 1547-6537. 48 (22-24), 1165–1213. Available from: https://www.tandf.co.uk/journals/titles/10643389.asp. doi:10.1080/10643389.2018.1518860.

Zainal, Z.A., Ali Rifau, G.A.Q., Seetharamu, K.N., 2002. Experimental investigation of a downdraft biomass gasifier. Biomass Bioenergy. 0961-9534. 23 (4), 283–289. doi:10.1016/s0961-9534(02)00059-4.

Zhang, W., Chen, J., Fang, H., Zhang, G., Zhu, Z., Xu, W., Mu, L., Zhu, Y., 2022. Simulation on co-gasification of bituminous coal and industrial sludge in a downdraft fixed bed gasifier coupling with sensible heat recovery, and potential application in sludge-to-energy. Energy. 0360-5442. 243, 123052. doi:10.1016/j.energy.2021.123052.

Zhang, X., Rajagopalan, K., Lei, H., Ruan, R., Sharma, B.K., 2017. An overview of a novel concept in biomass pyrolysis: microwave irradiation. Sustain. Energy Fuels. 1 (8), 1664–1699. doi:10.1039/c7se00254h.

CHAPTER 7

Life cycle assessment and sustainability of bioenergy from agricultural residues

Lakshmipriya Rajendran[a], Ninu Sunil[a], Aravind Madhavan[b], Reshmy Rajasekharan[c], and Raveendran Sindhu[a,d]
[a]Department of Food Technology, TKM Institute of Technology, Kollam, Kerala, India
[b]School of Biotechnology, Amrita Vishwa Vidyapeetham, Amritapuri, Kollam, Kerala, India
[c]Cellupro Green Pvt. Ltd., Industrial Development Area, Mavelikara, Kerala, India
[d]Santhigiri Scientific and Industrial Research Institute (SSIRI), Trivandrum, Kerala, India

7.1 Introduction

The world's two biggest problems right now are the rise in the need for food and energy on a worldwide scale. Because of the greenhouse effect caused by excessive CO_2 emissions, fossil fuels are not renewable and provide the majority of the energy in today's world. As a renewable resource with an abundant supply and the potential to reduce CO_2 emissions, using plant biomass as an energy source is a sustainable choice. Because they enhance crop value and do not compete with food, agricultural leftovers make intriguing feedstocks for bioenergy and may be easier to handle in many areas. There has been prior research on breeding crops to produce both food and bioenergy. However, the optimum plant characteristics vary depending on whether lignocellulosic wastes are fermented to produce biofuel or burned for heat or electricity. Keeping leaves green may be linked to increased stover moisture during harvest, which is unfavorable for thermochemical conversion. Climate change and the existence of diverse environmental conditions around the world make it imperative that the feedstock sources for the biofuel industry be diversified. This will open up new opportunities for the sector, greatly augmenting its fuel supply and enhancing its capacity to support political objectives related to energy security (Caicedo et al., 2016).

To generate value goods, promote sustainable development, and maximize ecological and socioeconomic benefits in the circular bioeconomy, it is imperative to strategically use resources for optimal biomass valorization. More value-added products can be produced with the help of emerging

Bioenergy from Agricultural Residues
https://doi.org/10.1016/B978-0-443-24726-2.00008-0

and enhanced technologies that maximize the utilization of plentiful biomass. Potential feedstocks include sugar beets, wheat, corn, surplus food, wood, sugarcane, agricultural residues, straw, and aquatic biomass. The bioeconomy includes the creation and conversion of bioresources and biowaste streams into value-added products, such as food, feeds, bioproducts, and bioenergy, in both established and emerging industries, such as foods, fisheries, agriculture, biotechnological, chemical, paper manufacturing, and energy-producing industries. Recent developments enable more efficient use of biomass, reduced processing costs, and more sustainable biomass utilization. It is possible to produce biofuels and organic fertilizers by using biomass waste in biorefineries. To facilitate the successful exploitation of bioresources, further study and development are needed to create efficient biorefining technologies (Awasthi et al., 2020).

7.2 Types of agricultural residues used for bioenergy

Agricultural residues are carbon-based compounds that are produced as a byproduct of agricultural crop harvesting and processing. During harvesting, primary or field-based residues are created; secondary or processed-based residues are produced along with the product during processing. Agricultural residues differ in bulk density, moisture content, particle size, and distribution based on operational treatment. They are frequently fibrous, low in nitrogen, and vary by area. These field wastes are sometimes used as fertilizer, erosion control, and calf feed. Almost half of these resources are burned on the farm before the start of the new farming season. Aside from the grains of crops used for food, such as corn, wheat, and rice, the residues or byproducts of these grains are also a valuable resource (Mohammed et al., 2018).

Energy crops include starch and sugar-rich crops such as maize and sugarcane, as well as oilseed crops such as soybeans and sunflowers. Sugar and starch crops are widely used as human food and animal feed. These crops and their byproducts can be easily transformed into biofuel, such as ethanol, using a simple fermentation process and then used as fuel. This category may also contain grass grown for hay, pasture, or soil protection. These crops are utilized as feedstock for energy production because they contain more fibers, such as cellulose, hemicellulose, and lignin, and fewer carbohydrates, proteins, and oils. These crops can generate energy in a variety of methods, including direct combustion for heat and/or power and ethanol synthesis from cellulose. Rice, wheat, and corn straw, as well as sugarcane bagasse, are widely regarded as important agricultural waste feedstocks for biofuel production.

7.2.1 Sugarcane bagasse

Sugarcane bagasse, the most significant agro-industrial residue, is a fibrous residue of cane stalks left over from the crushing and extraction of sugarcane juice. This byproduct of the sugar industry is mostly used as boiler fuel by sugar factories. Bagasse, in comparison to other agricultural leftovers, can be thought of as a rich solar energy reservoir due to its high yields and annual regeneration capability. Several procedures and products are now being described that use sugarcane bagasse as a raw material. This includes energy generation, pulp and paper manufacture, and a variety of fermentation-based products such as industrially essential enzymes, bioethanol, organic acids, alkaloids, protein-enriched bovine feed, antibiotics, and so on. Bagasse is typically used for cogeneration of heat and power, but it is also utilized for the manufacturing of building materials (Sindhu et al., 2016a).

7.2.2 Rice straw

Rice straw, because of its abundance all throughout the world, has enticing characteristic features. It is one of the most important crops, producing a substantial amount of silica-enriched biomass (Sharma et al., 2020). Rice straw is a visually appealing lignocellulosic material that can be fermented or gasified to produce liquid or gaseous fuel oils such as bioethanol, biogas, or methane. Furthermore, rice straw may be explored for the manufacture of biohydrogen, in which the straw is transformed to bio-oil via quick pyrolysis and the bio-oil is steam reformed to produce hydrogen. Finally, rice straw can be converted into a solid fuel that can be burned immediately or after further processing (Ibrahim, 2018).

7.2.3 Corn stover

For several reasons, including its abundance in the United States, reasonable homogeneity, and focused concentration for biofuel production, corn stover (*Zea mays*) is one of the most promising annual crops used as feedstocks for biofuels. If harvesting corn stover is not done in a sustainable way, it may have long-term implications on soil erosion and quality. The components of a corn stover include the stalk, leaves, sheaths, husks, shanks, cobs, tassels, lower ears, and silks (Pennington, 2020).

7.2.4 Wheat straw

Wheat straw is rich in minerals (calcium and phosphorus), proteins, silica, ash, and acid detergent fibers. It is also heavy in carbohydrates (lignin, cellulose, and hemicelluloses). Vitamins and bioactive compounds are also abundant in it. Because wheat straw is cheap, widely available, renewable, and safe for use in the environment, it is the perfect material for biorefineries to use to produce bioethanol, biogas, and biohydrogen. Wheat straw is a moiety that possesses antiinflammatory, antimicrobial, antiartherogenic, antiallergenic, antioxidant, antithrombotic, and other health effects (Tufail et al., 2021).

7.2.5 Impact on soil, water, and biodiversity

Land use conversion and the potential water consumption of bioenergy crops are the main causes of the effects of bioenergy production on water quantity. As the development of bioenergy crops increases, there is a major risk to water quality from nutrient pollution via surface runoff and infiltration into groundwater. Nitrate is the most significant polluting cause of nutrient contamination.

Ecological services and food production are both influenced by biodiversity. The initial land use scenario, the type of bioenergy production system, and the landscape configuration all affect how biodiversity is impacted by biofuel production. The most significant factor affecting biological abundance is land use conversion since it directly alters the production system and land use conditions, which are also impacted by planting location and plant type.

The most significant measure of soil quality is soil organic carbon (SOC), because high SOC levels support agricultural productivity, biodiversity, and soil water retention. There are three main ways that bioenergy production affects SOC: tillage, land use change, and residue removal. Harvesting dead plant remains from replanted croplands generally results in a decrease in carbon intake, which directly increases SOC loss. Appropriate residue management, such as limiting residue removal and additional organic matter inputs, can help reduce SOC loss to some extent.

7.3 LCA and its importance for sustainable energy production

According to Roy et al. (2009), life cycle assessment (LCA), is a tool that may be used to assess the environmental impact of a process, product, or activity across the course of its life cycle. Basic LCA principles include the "cradle to grave" analysis and the use of a functional unit to which all flows of mass and energy, use of resources and land, as well as potential impacts and likely interventions, are set in relation as quantitative measures to assess a product system's environmental aspects. LCA takes into account all stages of a product's life cycle, from raw material extraction and manufacturing to distribution, use, and end-of-life disposal. This holistic perspective helps in identifying and assessing the environmental impacts at every stage. LCA may help government and nongovernment organizations by implementing eco-labeling and environmental product declarations, and it can also help industry decision-makers by identifying possibilities to improve a product's environmental performance. LCA provides decision-makers with comprehensive information about the environmental performance of different products or processes. This information can be used to make informed decisions, whether in product design, process optimization, or policy development. LCA aligns with the principles of sustainable development by considering economic, environmental, and social aspects. It contributes to achieving goals related to responsible consumption and production, climate action, and environmental protection. A way to evaluate the sustainability of different energy sources is made possible by the growing need for sustainable renewable energy sources to lower pollution and rely less on traditional energy sources. A single renewable energy source might be very desirable for producing heat but less desirable for producing electricity and being used for transportation. Production of raw materials, technology, byproducts, and other factors are only a few of the many factors that need to be carefully taken into account in order to produce these energy sources on a commercial scale. Policymakers can select the most suitable energy source for a given purpose by using the LCA, an efficient tool for assessing the sustainability of different renewable energy sources. LCA is supported by international standards such as ISO 14040 and ISO 14044, providing a standardized methodology for conducting assessments. This ensures consistency and comparability of results across different studies (Pickel and Eigner, 2012).

7.4 LCA in bioenergy

Bioenergy is a very relevant alternative to renewable energy sources on a worldwide scale. Bioenergy, as opposed to fossil fuels, has the potential to be carbon neutral and is crucial in lowering greenhouse gas (GHG) emissions. Ninety percent of the ultimate biomass energy consumption is made up of biomass heat and electricity, and numerous studies of biofuel LCAs have been published. Nevertheless, the majority of these assessments concentrate on a limited number of impact categories, and very few of them address the production of heat and power from biomass. A comprehensive quantitative evaluation was not included in any examination of biomass electricity and heat LCAs. Research gaps in bioenergy LCA studies include the omission of the quantitative overview, the inadequate consideration of effect categories, and the failure to account for heat generation (Muench and Guenther, 2013). The LCA is a vital tool for calculating and assessing the environmental effect of a specific bioenergy system (Ghosh et al., 2020).

The several studies provided highlight the value of using a life cycle approach to evaluate how bioenergy systems would affect the environment. These LCA evaluations highlight areas for environmental improvement that are not discernible when concentrating solely on a single production process. All of the studies have one thing in common: they highlight the benefits of using biomass in agricultural systems appropriately. Though they aren't displayed here, comparable findings have been made in other systems, such as rice (e.g., using rice husk and straw). This means that if energy, environmental, and financial gains are to be realized, a focus on the sustainable use of biomass resources is necessary. But to actually put all of the byproducts to use, the right infrastructure would need to be established (power plants, biogas production facilities, fertilizer factories, etc.). Only then would these benefits become possible. It is essential to evaluate such systems using an integrated approach, such as LCA, to identify any trade-offs between the different life cycle stages (Gheewala, 2011).

7.5 LCA of GHGs from bioenergy production

LCA is a systematic, all-inclusive, and globally recognized approach to evaluate a product's environmental effects. It measures all pertinent emissions and resource consumption as well as the consequences on the environment, human health, and resource depletion. Numerous life cycle evaluations (LCAs) contrast traditional petroleum fuels and biofuels with regard to energy consumption and GHG emissions. A deeper understanding of the LCA concerns in the evaluation of GHG net emissions from biofuels has led to improved models and the hunt for better data (carbon stocks, indirect land use change (iLUC), coproducts treatment, and N_2O emissions) over the previous 5 years. This has drastically altered certain previous findings (e.g., iLUC estimates). The complexity of different feedstocks, locations, soils, local land use settings, and conversion processes necessitates more data and even better assessments in order to provide strong support for policies. Rising LCA-based reductions in GHG emissions and a gradual decline in land use change (LUC) emissions are reasonable expectations given the current trends in feedstock production and conversion efficiency, the increasing penetration of renewable energy sources into the energy matrix, the slowing rates of deforestation, and the tightening of land use and clearing regulations. It might be very important to consider when GHG emissions and removals occur, especially in forest-based biomass systems with extended cycle times. Remember that in certain cases, the emissions of aerosol and black carbon from biomass conversion systems, along with the albedo shift caused by LUC (for biomass production), can significantly contribute to global warming. Biofuels are deemed acceptable if they produce a discernible decrease in GHG emissions and minimize their adverse effects on the environment and society. Enhancing data collection is crucial for analyzing impacts. Tools for national (GBEP 2011) and project-level use (GEF 2013; ISO 2013) have been developed to assist public policies (planning and monitoring) (de Carvalho Macedo et al., 2015).

7.6 Conclusion

This chapter underscores the pivotal role of LCA in evaluating the sustainability of bioenergy from agricultural residues. Through a holistic examination of environmental, social, and economic factors, it becomes evident that responsible bioenergy production necessitates a nuanced understanding of its entire life cycle. Emphasizing the need for continuous innovation and integration of sustainable practices, the chapter advocates for a balanced approach that not only meets energy demands but also ensures the preservation of ecosystems and the well-being of communities. As we navigate the dynamic landscape of renewable energy, the insights gained from this analysis serve as a compass for steering bioenergy initiatives towards a truly sustainable and resilient future.

References

Awasthi, M.K., Sarsaiya, S., Patel, A., Juneja, A., Singh, R.P., Yan, B., Taherzadeh, M.J., 2020. Refining biomass residues for sustainable energy and bio-products: an assessment of technology, its importance, and strategic applications in circular bio-economy. Renew. Sustain. Energy Rev. 127, 109876.

Caicedo, M., Barros, J., Ordás, B., 2016. Redefining agricultural residues as bioenergy feedstocks. Materials 9 (8), 635.

de Carvalho Macedo, I., Nassar, A.M., Cowiec, A.L., Seabra, J.E., Marellid, L., 2015. Greenhouse gas emissions from bioenergy. Bioenergy Sustain. Bridging Gaps SCOPE 72, 582–617.

Gheewala, S.H., 2011. Life cycle assessment (LCA) to evaluate environmental impacts of bioenergy projects. J. Sustain. Energy Environ. Spec. Iss. 35 (1), 35–38.

Ghosh, P., Sengupta, S., Singh, L., Sahay, A., 2020. Life cycle assessment of waste-to-bioenergy processes: a review. Bioreactors, 105–122.

Ibrahim, H., 2018. Bio-energy production from rice straw: a review. Recent Adv. Petrochem. Sci 5, 1–7.

Mohammed, N.I., Kabbashi, N., Alade, A., 2018. Significance of agricultural residues in sustainable biofuel development. In: Agricultural Waste and Residues. IntechOpen, p. 71–88.

Muench, S., Guenther, E., 2013. A systematic review of bioenergy life cycle assessments. Appl. Energy 112, 257–273.

Pennington, D., 2020. Bioenergy crops. In: Anju Dahiya (Ed.) Biomass to Biofuels and Waste to Energy. Elsevier, United States, Academic Press. pp. 133–155. Available from https://www.elsevier.com/books/bioenergy/dahiya/978-0-12-815497-7. doi:10.1016/B978-0-12-815497-7.00007-5.

Pickel, P., Eigner, M., 2012. Life cycle assessment (LCA) and its importance for the agricultural sector. In: CLub of Bologna-XXIII. Members' Meeting-Bologna (Italy). EIMA International, Bologna.

Roy, P., Nei, D., Orikasa, T., Xu, Q., Okadome, H., Nakamura, N., Shiina, T., 2009. A review of life cycle assessment (LCA) on some food products. J. Food Eng. 90 (1), 1–10.

Sharma, A., Singh, G., Arya, S.K., 2020. Biofuel from rice straw. J. Clean. Prod. 277, 124101.

Sindhu, R., Gnansounou, E., Binod, P., Pandey, A., 2016. Bioconversion of sugarcane crop residue for value added products—an overview. Renew. Energy 98, 203–215.

Tufail, T., Saeed, F., Afzaal, M., Ain, H.B.U., Gilani, S.A., Hussain, M., Anjum, F.M., 2021. Wheat straw: a natural remedy against different maladies. Food Sci. Nutr. 9 (4), 2335–2344.

CHAPTER 8

Policy and economic considerations for bioenergy from agricultural residues

Gurunathan Baskar[a], Ravichandran Pravin[a], Samuel Lalthazuala Rokhum[b], and Navnit Kumar Ramamoorthy[c]
[a]Department of Biotechnology, St. Joseph's College of Engineering, Chennai, Tamil Nadu, India
[b]Department of Chemistry, National Institute of Technology Silchar, Silchar, Assam, India
[c]Department of Biotechnology, School of Lifesciences, Central University of Rajasthan, Ajmer, Rajasthan, India

8.1 Introduction

The demand for crude oil has created a sudden rise in the cost of fossil fuels along with the concerns associated with environmental issues. To overcome these difficulties, there is a sudden need to transform into biofuels, which are normally produced from biomass resources. The ideology of "make-use-dispose" was not considered sustainable owing to the rise in population, which may increase up to 9 billion by the year 2050. It is also predicted by the IRENA renewable energy roadmap that if the conversion technologies as of now are implemented, the overall share of biomass's global energy share could be 36% by the year 2030 (Errera et al., 2023). Also, the total demand for the biomass from countries such as the USA, India, China, and Brazil could account for 56% in the year 2030. The concept of reuse or recycling has to be effectively implemented to face the upcoming challenges in the bioenergy sector around the world (Rupesh et al., 2024). Biomass is nothing but the organic substance that is produced directly or indirectly through the process of photosynthesis. Some examples of biomass are sugarcane bagasse, rice, and wheat straw, wood and municipal solid wastes (Adamu et al., 2023).

Biomass has been categorized into four generations. The first generation of biomass includes the crop plants where the debate between food versus fuel has shifted towards the use of lignocellulosic biomass as second generation. The challenges related to biomass utilization are the unavailability of the source throughout all the regions in the world, problems in transporting

Bioenergy from Agricultural Residues
https://doi.org/10.1016/B978-0-443-24726-2.00005-5

and storing the biomass, and basic conversion technologies and industrial chains. In view of these criteria, agricultural residues or waste has been considered to be the best alternative resource for the production of bioenergy in the replacement of fossil fuels (Mujtaba et al., 2023). While considering the sustainable production of biofuels, agricultural residue serves as a suitable and environmentally friendly source for the production of bioenergy. The research has been at full speed for the development of technologies to produce biofuels from agricultural residues. There are several research or review articles published on these criteria. Also, the opening of small-scale industries has been made for biofuel production in recent times. Although the resource availability and process technologies have been developed, there are still problems in the commercialization of the products. The main criteria fall in the implementation of proper policies in view of promoting biofuels production across the world (Sidana and Yadav, 2022).

This chapter mainly focuses on the utilization of agricultural residues for bioenergy production and their challenges in commercialization. It also explores the policies and measures to be taken for the sustainability of bioenergy production worldwide. The United Nations Sustainability Development Goals (SDG), which correspond to biofuel production from agricultural waste were discussed. Also, the current challenges and future perspectives for the sustainable production of biofuels from agricultural waste have been discussed.

8.2 Agricultural residue and its composition

Agriculture is considered to be one of the most commonly practiced activities all over the world. The report by FAO says that nearly 1.5 billion hectares have been used for agriculture, and this could increase up to 2.7 billion hectares in due time because of the increase in population. Cereals such as maize, barley, millets, and wheat were cultivated majorly; among them, rice, wheat, and maize were grown on nearly two-thirds of the global agricultural lands. Other crops such as sugarcane, potatoes, palm, canola, sunflower, and rapeseed were also grown across the world. To do this, numerous wastes have been generated during the processing of these crops that are thrown away in the empty lands, which causes pollution to land and water. Also, burning of these wastes in huge amounts created an urgent consideration to lower air pollution. To eradicate all these, the utilization of this agricultural waste can be efficiently involved in the production of biofuels and bioenergy (Saravanan et al., 2021).

The agricultural residue has been generated from four different categories, such as residues from crops, waste generated from processing industries, waste generated during the consumption of food, and livestock waste. The waste can be collected and could be used for the production of various bioenergy and value-added products. The part of the crop residues generated from wheat, rice, oats, and sorghum was used as fodder, while most of them were burned in empty lands. The residues such as the peels of fruits and vegetables and pomace were thrown away after the extraction of juice, and de-oiled cakes were generated from the food processing industries. Among them, sugarcane bagasse and oilseed cakes were considered to be the most generated waste that needs to be utilized. Due to the increase in consumption of meat, livestock waste has been generated in numerous amounts. It is reported that nearly 56 billion of livestock were grown to meet the demands of people and this could double in number by the year 2025. Therefore, the proper conversion of these wastes to bioenergy is of the utmost consideration (Kumar et al., 2023).

The agricultural residues majorly consist of lignin, cellulose, hemicellulose; hence, they are also called lignocellulosic wastes. The cellulose content of around 30%–50%, followed by hemicellulose 20%–38% and lignin of 7%–21%, was found in the agricultural residues. The composition of some of the sources has been tabulated in Table 8.1. The carbohydrate content of nearly 40%–85% will be present in the wastes that are generated from processing plants. These carbohydrates can be used for bio-alcohol and biogas production through biochemical processes. Therefore, the use of these wastes for bioenergy production helps in achieving the concept of minimization of waste with a cleaner and sustainable environment (Kulisic et al., 2021).

Table 8.1 Composition of agricultural waste sources.

	Composition of the source		
Agricultural waste sources	**Lignin (%)**	**Cellulose (%)**	**Hemicellulose (%)**
Sugarcane bagasse	11	43	31
Rice straw	13	31	22
Banana stems	8	60	8
Wheat straw	19	36	24
Corn straw	20	43	21
Oat straw	20	40	27

8.3 Biofuels and bioenergy production from agricultural residue

The agricultural residue can be transformed into several products such as biodiesel, bioethanol, biogas, organic acids, enzymes, and biofertilizers. The methods such as thermochemical and biochemical processes have been used for the conversion of these wastes into usable products. Based on the raw material used, the selection of conversion process is decided to obtain the possible highest yield of the product. In view of biochemical conversion, the raw materials with a higher amount of cellulose and hemicellulose and a high C/N ratio were selected. While the thermochemical conversion chooses the raw materials with higher lignin content (Hoang et al., 2022). The most widely produced products from the utilization of agricultural waste have been discussed below and shown in Fig. 8.1.

8.3.1 Bio-oil

Bio-oil is a liquid fuel derived from lignocellulosic agricultural residues through thermochemical conversion processes such as pyrolysis and gasification. Gasification involves the partial combustion of biomass at high temperatures above 700°C in the presence of air or steam, producing syngas, which can be further processed into liquid fuels (Hameed et al., 2020). Pyrolysis, on the other hand, is a thermal decomposition process carried out at temperatures ranging from 300°C to 700°C in the absence

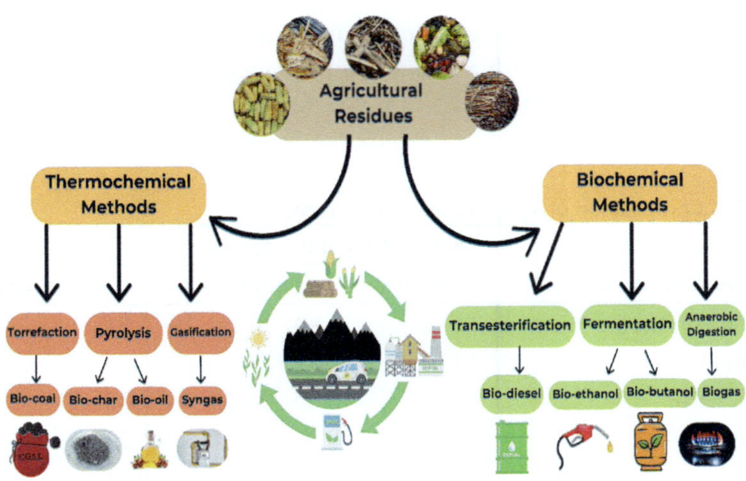

Fig. 8.1 Production of various biofuels and bioproducts from agricultural residues through different methods.

of oxygen. Among these methods, fast pyrolysis is widely used for bio-oil production due to its ability to yield higher liquid fractions.

Despite its potential as a cleaner alternative to conventional fossil fuels, bio-oil possesses certain drawbacks that limit its direct use as a transportation fuel. These include high viscosity, high oxygen and moisture content, corrosiveness, and instability, which contribute to lower energy density compared to traditional diesel or gasoline. As a result, bio-oil often requires upgrading or blending with petroleum-based fuels before its utilization. Advanced refining techniques, such as hydrodeoxygenation, catalytic cracking, and emulsification, have been explored to improve the fuel properties of bio-oil, making it suitable for applications in transportation, power generation, and industrial heating. Wood waste, agricultural residues, and animal manure serve as common feedstocks for pyrolysis-based bio-oil production, offering a sustainable pathway for waste valorization while reducing reliance on fossil fuels (Sharma et al., 2023a).

8.3.2 Biochar

Biochar, derived from agricultural residues through pyrolysis, has gained interest as a valuable byproduct with multiple applications in bioenergy production. The process of pyrolysis, conducted at temperatures ranging from 300°C to 700°C in an oxygen-limited environment, converts biomass into biochar, bio-oil, and syngas. Among these products, biochar stands out due to its high carbon content, porous structure, and potential for enhancing bioenergy processes (Pravin and Baskar, 2024). Agricultural residues such as rice husks, banana peels, peanut shells, and cotton stalks serve as excellent feedstocks for biochar production. The use of biochar in bioenergy systems extends beyond direct combustion; it plays a crucial role in improving the efficiency of anaerobic digestion, biomass gasification, and microbial fuel cells. For instance, the addition of biochar to anaerobic digesters enhances microbial activity, leading to higher biogas yields. Similarly in biomass gasification, biochar contributes to improved syngas quality by acting as a tar-reducing agent. These applications demonstrate the versatility of biochar as a means to optimize bioenergy generation while simultaneously promoting waste valorization (Low and Yee, 2021).

Furthermore, biochar improves soil properties when used as a soil amendment, thereby supporting biomass production for future bioenergy applications. The presence of essential minerals in biochar enhances nutrient retention, reduces soil acidity, and increases water-holding capacity, ultimately leading to higher agricultural productivity. By integrating biochar into bioenergy strategies, the sustainability of agricultural residue-based energy systems can be significantly improved, ensuring both economic and environmental benefits. The development of biochar-based technologies aligns with circular economy principles, where agricultural waste is repurposed to enhance renewable energy production while minimizing environmental impact (Anand et al., 2021).

8.3.3 Biodiesel

Biodiesel, which is also known as fatty acid methyl esters, is mainly produced from extracted oils. Biodiesel is considered to have the same properties that conventional diesel possess with the advantages of lower carbon emissions, nontoxicity, and biodegradable fuel. The transesterification method is widely used for the production of biodiesel, where the fatty acids along with the alcohol in the presence of a catalyst have been converted to methyl esters and glycerol (Sangeetha et al., 2024). The advancements in recyclable catalysts, which are produced again from agricultural waste, add an extra benefit to the process. The waste cooking oils and oils generated from non-edible seeds were considered to play an important role in revolutionizing the biodiesel market due to their advantages such as the lower dependence on edible oils and higher consideration of the food chain. The cost also plays a major role in promoting the use of non-edible and waste oils for the biodiesel production which can be obtained at cheaper rates (Ravichandran et al., 2023).

8.3.4 Bioethanol and biobutanol

Wastes such as wheat and rice straw and sugarcane bagasse were efficiently used for the production of bioethanol. Bioethanol has several advantages over petroleum such as a higher octane number and carbon content. Biobutanol can also be used as an alternative fuel because of its lower volatility and vapor pressure and higher energy content. Both bioethanol and biobutanol were produced using the fermentation process. Fermentation is a biochemical process where the sugars have been used as substrates for the production of biofuels. However, materials such as lignocellulosic waste need a prior step of enzymatic hydrolysis where the complex sugars are broken down to simpler forms using the enzyme (Elumalai et al., 2023). Both the enzymatic hydrolysis and fermentation processes can be done sequentially or concurrently, known as separate hydrolysis and fermentation and simultaneous saccharification and fermentation, respectively. Also, byproducts such as hydrogen gas and acids such as acetic, lactic, and butyric can be produced. But these byproducts can be toxic to the microorganisms, which makes the process difficult. To overcome this, microorganisms with less toxicity and higher yielding efficiency should be preferred (Adekunle et al., 2016).

8.3.5 Biohydrogen

The cellulose and hemicellulose content of the agricultural residues was used for the production of biohydrogen through autotrophic or heterotrophic transformation. Biohydrogen has gained a lot of attention due to its advantage of having no carbon content, which creates only water as a byproduct, unlike combustion (Deivayanai et al., 2022). However, the complicated structure of lignocellulosic materials needs prior treatment for the microorganisms to enhance the hydrolysis of the cellulose. In terms of biohydrogen production, crop residues were preferred when compared to livestock waste since they contain nitrogen, which may produce a higher amount of ammonia, leading to failure of the reactor. Biohydrogen can be considered as a best alternative fuel because of its several advantages to environmental sustainability and wider usability (Khan et al., 2023).

8.4 Biofuels policy and subsidies for sustainable bioenergy production in various countries

In the view of higher energy demands and environmental problems with the use of petroleum fuels, the necessity to transform to biofuels produced from renewable and economically feasible feedstocks has been gaining attention. Although the research on biofuels has provided positive pathways and ideas on large-scale production, the commercialization of the biofuels becomes the biggest question in the current stage. To overcome this, the implementation of necessary policies, frameworks, and subsidies is needed for promoting the growth of biofuels in the market (Usmani et al., 2023).

8.4.1 India

The first biofuel act was introduced in India in the form of India's Power Alcohol Act in the year 1948, which was not fully focused on and achieved due to the lack of feedstocks and technologies. Later in 2009, India announced the National Biofuel Policy, which targets the blending of 20% ethanol in petrol and 5% blending of biodiesel in diesel fuel, respectively. The announcement of these policies has led to an increase in biofuel production, which helped to attain nearly 9.3% of ethanol blending in petrol fuels. To achieve the targetable 20% blending, the country has made some amendments in the year 2022 to reduce the timeframe of achieving these targets from 2030 to 2025 (International Energy Agency (IEA), 2023). Due to the higher logistics price of feedstocks, the biodiesel price is not considered to be economically viable. India's biofuel policy has announced only minimum incentives to producers and none to consumers. This led to the problem of attaining the blending targets, which rose to the revision of the policy in 2022. This new policy has introduced favorable support to the oil marketing companies to attain the ethanol production possibly. The government of India has funded 41.5 million dollars for the sugar mills for the enhancement of the technologies for biofuel production. To do this, the government also reduced the export subsidy from 82 to 55 dollars per metric ton. Even though India generates nearly 2.2 billion liters of waste cooking oil, which can be converted to biodiesel, the country has not introduced any new plans for the promotion of biodiesel production in the country. To promote the cellulosic ethanol production, the government has funded nearly 5000 crores and guided the marketing companies to sign the agreement for the purchase of ethanol for 15 years from the producers. These initiatives can promote biofuel production in the country and will definitely help the country to achieve its target of blending in due years (Saravanan et al., 2020).

8.4.2 Brazil

The most needed goal in promoting the biofuels is blending them with the conventional fossil fuels and using them in the transportation sector. In view of that, Brazil is considered to have the highest blending rates of biofuels and is in second place in global biofuel utilization. The PROALCOOL program was started by Brazil in the year 1975 to promote the replacement of gasoline with ethanol. This program was created due to the oil crisis in 1973, and the PROALCOOL program was a huge success, which created a huge revolution in the bioethanol sectors of Brazil. The subsidies and tax incentives were provided to promote the flex-fuel vehicles (FFV), and it is known that now 98% of the total cars sold in Brazil are FFV. After the success of this program, Brazil has also introduced another program named PRO-BIODIESEL in the year 2004. The blending of biofuels started at 2%, which reached up to 20% in 2020. Also, the manufacturing industries are instructed to purchase raw materials for biodiesel production directly from the farmers without intermediaries to promote the concept of "Social fuel". Brazil has also announced a new policy called Reno-vaBio in the year 2017 to reduce the emissions by 37% by 2025. The government of Brazil has also not implemented any tax changes on bioethanol. The contribution for intervention in the economic domain was fixed at zero, and the tax for circulation of goods and services depends on the total revenue of the government (Grangeia et al., 2021).

8.4.3 China

China is the 4th largest country in ethanol production and 7th largest country in biodiesel production around the world. Even though the production rate is higher, they are lagging in achieving their blending rates properly. China's ethanol production was 40 million liters but only able to achieve 2% of the blending rate in the year 2021. The increase in the corn prices has created a downfall in bioethanol production (Song et al., 2022). The corn price is the main criterion behind China's biofuel policy. Recently China has also introduced policies related to biodiesel production; however, due to their absence of well-developed technologies, it is not considered as important as ethanol. In the beginning stages, China provided full tax exemption and also provided subsidies for bioethanol production through corn but this has been stopped completely in 2019 due to an increase in food prices. In view of this, they are now focusing on cellulosic bioethanol, which is majorly produced from agricultural residues. However, the cost competition over crude oil has failed in the success of cellulosic bioethanol commercialization. China needs to implement new support policies for consumers to achieve their blending mandates efficiently. China has started an ethanol production plant in Heilongjiang's Hailun city with a 30,000 liter capacity in the year 2020. Due to the success of this plant, the country has also planned to start a few more production plants in Jilin province (Zhao, 2015).

8.4.4 European Union

The European Union's biofuel policy has already introduced several regulatory policies and subsidies to promote biofuel production. They have also focused more on the reduction of carbon emissions in their policy. The first biofuel policy was implemented in the year 2003, stating the goal to attain a 5.75% share in the transport sector from biofuels by the year 2010. Later in 2011, they achieved a share of 3% in the transport sector (Hassan et al., 2018). The production and consumption of biofuels in the European Union have increased in recent times after the outbreak of COVID and Russia-Ukraine war issues. The higher availability of sugar beets, wheat, and corn kernels has been a better environment for the increase of ethanol production in the country. The European Union has introduced several incentives such as tax exemptions for the promotion of biofuels after the approval from the union. The incentives can be offered for six years and extended if needed (Pearson and Turner, 2014).

8.4.5 United States of America (USA)

The United States took significant steps toward integrating biofuels into its energy policy in response to the 1970s oil crisis. Since then, the country has implemented strong policy measures to support biofuel production. In 2022, the United States produced approximately 52 billion liters of ethanol, contributing to 10% of its gasoline usage and accounting for 55% of global ethanol output, securing its position as the world's leading ethanol producer. In terms of diesel consumption, the country utilizes around 178 billion liters annually, of which biodiesel contributes about 15 billion liters fulfilling nearly 9% of the national diesel demand. The United States government initially introduced financial incentives for biofuels under the Energy Tax Act of 1978, granting a tax exemption of $0.04 per gallon for ethanol-blended fuel. This initiative was later reinforced by the American Jobs Creation Act of 2004, which was subsequently extended under the Energy Policy Acts of 2005 and 2010. These acts introduced the Volumetric Ethanol Excise Tax Credit, offering ethanol blenders and retailers a tax incentive of $0.45 per gallon. However, this credit was discontinued in December 2011. In addition, the Farm Bill of 2007 modified tax benefits for ethanol, reducing incentives for first-generation ethanol from $0.51 to $0.45 per gallon while introducing a higher incentive of $1.01 per gallon for cellulosic ethanol (Su et al., 2015).

Recognizing the long-term challenges associated with food-based biofuels, the United States gradually transitioned its focus to advanced biofuels. The Renewable Fuel Standard set specific mandates for advanced biofuel production to ensure sustainability. Financial support for research and development in this sector is substantial, with various funding initiatives backing innovation. The Biomass Research and Development Act (BRDA) of 2000 allocated $49 million to foster collaboration between the United States Department of Energy (USDOE) and the United States Department of Agriculture (USDA). The Energy Policy Act of 2005 further boosted BRDA funding by $500 million to accelerate research in biofuel feedstock development.

Furthermore, the United States government has facilitated strong collaborations between public and private institutions to advance biofuel technology. Leading research organizations such as the National Renewable Energy Laboratory, Oak Ridge National Laboratory, and RAND Corporation, along with agencies such as the USDOE and USDA, play a pivotal role in biofuel innovation. These efforts have strengthened the country's position in sustainable bioenergy development (Usmani et al., 2023).

8.4.6 Canada

Canada introduced nationwide biofuel blending mandates in 2010, requiring a 5% ethanol blend in gasoline and 2% biodiesel in diesel. Since then, the country has witnessed a steady rise in biofuel consumption surpassing its initial blending targets. By 2020, ethanol consumption had reached 2.7 billion liters, contributing around 6% of gasoline demand. However, biodiesel consumption has remained relatively stable over the years, with a modest increase in 2020 raising the national blend rate slightly above 3%. Canada's biofuel policies are closely aligned with its climate goals. In 2020, the government introduced the Clean Fuel Standard Draft-1, outlining strategies to lower the carbon intensity of fossil fuels. Key measures included higher carbon taxes, regulatory frameworks for net-zero emission vehicles, and a mandate requiring all new light-duty vehicle sales to be zero-emission by 2035. These policies reflect Canada's commitment to addressing carbon emissions through bioenergy initiatives.

Financial incentives play a crucial role in Canada's biofuel development. The ecoAUTO Rebate Program, launched in 2007, offered consumers rebates between $1000 and $2000 for purchasing or leasing fuel-efficient vehicles. In 2008, the government allocated $1.48 billion in seed funding to support domestic biofuel producers. Further investments included the National Biomass Expansion Program, which dedicated $140 million to new ethanol plant construction, and the ecoAgriculture Biofuels Capital Initiative, which provided $200 million between 2011 and 2015 for the development of advanced biofuel production facilities. More recently, in 2021, the Clean Fuels Fund was launched, committing CDN $1.5 billion over 5 years to enhance zero-emission fuel production and distribution. Canada has also made notable progress in scaling up advanced biofuel production. Enerkem, a leading biofuel producer, commenced cellulosic ethanol production in 2012 using biochemical conversion processes. Their initial facility had a 5 million liter capacity and utilized treated wood waste as feedstock. The company later expanded with a 38 million liter cellulosic ethanol plant in Edmonton, Alberta, which processes 1 million dry metric tons of municipal solid waste annually. These advancements underscore Canada's efforts in transitioning towards a more sustainable bioenergy landscape (Wolinetz and Hein, 2017).

8.4.7 Japan

Japan's commitment to renewable energy is driven by mainly two primary factors, which are environmental protection and reducing dependence on energy imports. Following the Fukushima disaster, the country's energy import costs surged to $30 billion, prompting the government to incorporate renewables into its Fundamental Energy Plan. The national target aims for over 20% of power generation to come from renewable sources by 2030. Bioethanol consumption in Japan has been steadily increasing. By 2020, the country consumed 850 million liters of ethanol, accounting for approximately 2% of total gasoline consumption. However, biodiesel adoption remains minimal, contributing only a negligible share of total diesel use.

To incentivize biofuel adoption, the Japanese government offers various tax benefits. In addition, the Law for Feedstock Promotion, introduced in 2008, provided tax breaks and financial support to both biofuel producers and feedstock suppliers. Japan actively invests in advanced biofuel research, particularly in microalgae-based biofuels. In 2010, the Ministry of Agriculture, Forestry, and Fisheries initiated a research program to explore oil extraction from the microalgae through public-private partnerships. Moreover, Japan has engaged in international collaborations to develop low-emission biofuel production technologies. In 2011, the Japan Science and Technology Agency partnered with the United States National Science Foundation, funding four joint research teams with a $12 million grant to explore efficient algae-based biofuel production (Oyama and Masutani, 2017).

8.4.8 Australia

Australia has a sufficient amount of biomass for their biofuel production but only relies on the import of fuels for their usage. Only two states, namely New South Wales and Queensland have announced their blending mandates in Australia. The biodiesel production rate was good in Australia but seems to be reduced in the year 2016. To encourage bioenergy production, the government has introduced several programs and grants, but the lack of introduction of a blending policy is still present. To promote the sustainable production of biofuels in Australia, the Clean Energy Finance Corporation has announced it will invest nearly 10 billion Australian dollars. The Australian Centre for Renewable Energy invested about 690 million dollars for the promotion of renewable energy. The government has invested in several biofuel projects to promote the production all over the country (Ogunkunle and Ahmed, 2019).

8.4.9 Malaysia

As one of the world's leading producers and exporters of palm oil, Malaysia primarily focuses on biodiesel production rather than fuel ethanol due to the lack of a domestic ethanol feedstock supply. The country's vast palm oil reserves play a crucial role in its carbon reduction commitments and SDGs. Malaysia adopted its National Biofuel Policy in 2006, aiming to reduce reliance on imported fuels and promote eco-friendly, renewable energy sources. This was followed by the enactment of the Malaysian Biofuel Industry Act (2007), which took effect in 2008, leading to an initial 5% biodiesel blend requirement for government vehicles. Under the Eleventh Malaysia Plan (2016–2020), the biodiesel blending mandate was intended to gradually increase to 20% by 2020. However, the high cost of crude palm oil made maintaining the 20% blend rate economically unfeasible. The situation was further impacted by the COVID-19 pandemic, which significantly reduced global biofuel consumption. As a result, Malaysia revised its biodiesel blending mandate to 7% (Jayed et al., 2010).

To encourage biodiesel production, Malaysia offers investment tax allowances under the Investment Promotion Act (1986). Biodiesel producers classified under pioneer industry status are eligible for a 70% tax exemption on income from biodiesel production for 5 years. To counteract fluctuations in diesel prices, the Malaysian government employs the Automatic Pricing Mechanism, which provides subsidies to keep biodiesel economically viable (Masjuki et al., 2013).

8.5 Sustainable development goals

The utilization of agricultural residues for bioenergy production aligned with achieving multiple SDGs by promoting sustainable energy, climate action, responsible waste management, and economic growth (Mehmood, 2021). Agricultural residues such as rice husks, wheat straw, corn stover, and sugarcane bagasse serve as abundant, low-cost feedstocks for bioenergy production, aligning with SDG 7 (Affordable and Clean Energy). Their conversion into biofuels helps reduce dependence on fossil fuels, providing an eco-friendly and renewable energy alternative. In addition, biofuel production from crop residues supports SDG 13 (Climate Action) by mitigating greenhouse gas emissions, particularly by reducing open-field burning, which is a significant contributor to air pollution and environmental degradation.

The responsible utilization of agricultural waste for bioenergy also contributes to SDG 12 (Responsible Consumption and Production) by promoting waste valorization and reducing environmental pollution. Instead of allowing residues to decompose or be burned inefficiently, their conversion into bioethanol, biogas, and biochar adds economic and environmental value. The expansion of the bioenergy sector, particularly in rural areas, aligns with SDG 8 (Decent Work and Economic Growth) by creating employment opportunities across the supply chain, from feedstock collection and processing to biofuel distribution and research. Furthermore, investment in advanced biofuel technologies such as enzymatic hydrolysis, thermochemical conversion, and anaerobic digestion supports SDG 9 (Industry, Innovation, and Infrastructure) by improving process efficiency and developing sustainable energy infrastructure.

Despite these advantages, several challenges must be addressed to ensure the long-term sustainability of agricultural residue-based bioenergy. Issues such as feedstock logistics, seasonal availability, and the economic feasibility of large-scale production require targeted policy interventions and advancements in technology. By integrating sustainable practices with economic incentives, bioenergy from agricultural residues can significantly contribute to global sustainability goals while reducing the environmental impact of conventional energy sources (Das et al., 2023).

8.6 Economic consideration for sustainable bioenergy production

The economic feasibility of bioenergy production from agricultural residues depends on multiple factors, including feedstock availability, processing costs, logistics, government policies, and market demand. Agricultural residues, often considered waste, present a cost-effective alternative to primary feedstocks such as food crops. Since these residues are byproducts of existing agricultural activities, their use for bioenergy production can enhance rural incomes, improve energy security, and contribute to environmental sustainability. However, challenges such as high collection costs, storage issues, and supply chain inefficiencies must be addressed to make residue-based biofuels economically viable (Bui et al., 2022). The overall biofuel supply chain was shown in Fig. 8.2.

One of the primary economic advantages of agricultural residues is their relatively low cost compared to conventional biofuel feedstocks. However, their scattered nature and seasonal availability lead to high collection and transportation expenses, which can contribute up to 40% of the total biofuel production cost. The high moisture content of residues also necessitates appropriate drying and storage facilities to prevent degradation. Decentralized biorefineries, located near agricultural hubs, could be a practical solution to reduce transportation costs and improve economic efficiency. In addition, advancements in conversion technologies have improved the overall cost-effectiveness of biofuel production from residues. Technoeconomic analysis studies suggest that integrating multiple biofuels and bioproducts production can enhance profitability by utilizing all components of the feedstock (Sharma et al., 2023b).

Government policies play a crucial role in determining the economic viability of bioenergy from agricultural residues. Many countries offer financial incentives, including subsidies, tax benefits, and feed-in tariffs, to promote biofuel adoption. Policies such as E20 (20% ethanol blending) and B10 (10% biodiesel blending) create market stability and ensure consistent demand for biofuels derived from agricultural residues. Carbon credit programs and emissions reduction initiatives further enhance the financial attractiveness of residue-based biofuels, making them competitive with fossil fuels (Shabbir et al., 2023).

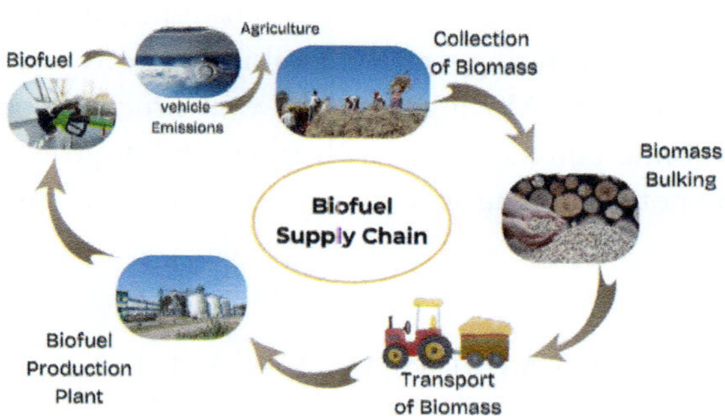

Fig. 8.2 Overall supply chain of biofuel production from agricultural residues.

Market dynamics also influence the economic competitiveness of bioenergy. The price of biofuels fluctuates based on global crude oil prices, government policies, and technological advancements. In addition, value-added products such as biochar, biofertilizers, and biochemicals contribute to the overall economic sustainability of biorefineries by providing additional revenue streams. The emerging markets for bioplastics, biobased chemicals, and green hydrogen derived from agricultural residues further expand the potential of this sector. Beyond economic viability, environmental and economic trade-offs must be considered. Life cycle assessment studies indicate that biofuels from agricultural residues have a significantly lower carbon footprint compared to first-generation biofuels derived from food crops. Since agricultural residues do not require additional land use, their conversion into bioenergy is both environmentally and economically sustainable. By integrating bioenergy production with sustainable agricultural practices, a circular economy model can be established, reducing reliance on synthetic fertilizers and fossil fuels.

8.7 Conclusions

Agricultural residues hold significant potential as a sustainable feedstock for bioenergy production, contributing to both energy security and environmental sustainability. Effective policy frameworks that support residue utilization can enhance rural economies, reduce dependency on fossil fuels, and mitigate waste management issues. The economic viability of bioenergy from agricultural residues largely depends on cost-effective collection systems, advancements in conversion technologies, and financial incentives that encourage large-scale adoption. While progress has been made in integrating bioenergy policies with national renewable energy goals, challenges such as fluctuating biomass availability, logistical constraints, and market competitiveness still need to be addressed. Strengthening regulatory mechanisms, investing in biorefinery infrastructure, and fostering technological innovations will be essential in overcoming these barriers. Moving forward, a well-structured policy approach that aligns economic considerations with sustainability goals will be key to maximizing the potential of agricultural residues for bioenergy production.

References

Adamu, H., Bello, U., Yuguda, A.U., Tafida, U.I., Jalam, A.M., Sabo, A., Qamar, M., 2023. Production processes, techno-economic and policy challenges of bioenergy production from fruit and vegetable wastes. Renew. Sustain. Energy Rev. 18790690. 186. Available from: https://www.journals.elsevier.com/renewable-and-sustainable-energy-reviews. doi:10.1016/j.rser.2023.113686.

Adekunle, A., Orsat, V., Raghavan, V., 2016. Lignocellulosic bioethanol: a review and design conceptualization study of production from cassava peels. Renew. Sustain. Energy Rev. 13640321. 64, 518–530. doi:10.1016/j.rser.2016.06.064.

Anand, A., Kumar, V., Kaushal, P., 2021. Biochar and its twin benefits: crop residue management and climate change mitigation in India. Renew. Sustain. Energy Rev. 156, 111959. doi:10.1016/j.rser.2021.111959.

Bui, V.D., Vu, H.P., Nguyen, H.P., Duong, X.Q., Nguyen, D.T., Pham, M.T., Nguyen, P.Q.P., 2022. Techno-economic assessment and logistics management of biomass in the conversion progress to bioenergy. Sustain. Energy Technol. Assess. 55, 102991. doi:10.1016/j.seta.2022.102991.

Das, P., Jha, C.K., Saxena, S., Ghosh, R.K., 2023. Can biofuels help achieve sustainable development goals in India? A systematic review. Renew. Sustain. Energy Rev. 192, 114246. doi:10.1016/j.rser.2023.114246.

Deivayanai, V.C., Yaashikaa, P.R., Kumar, P.S., Rangasamy, G., 2022. A comprehensive review on the biological conversion of lignocellulosic biomass into hydrogen: pretreatment strategy, technology advances and perspectives. Bioresour. Technol. 09608524. 365, 128166. doi:10.1016/j.biortech.2022.128166.

Elumalai, R.S., Ramanujam, P., Tawfik, M.A., Ravichandran, P., Gurunathan, B., 2023. Optimization and kinetics modelling for enhancing the bioethanol production from banana peduncle using *Trichoderma reesei* and *Kluveromyces marxianus* by co-pretreatment methods. Sustain. Energy Technol. Assess. 56, 103129. doi:10.1016/j.seta.2023.103129.

Errera, M.R., Dias, T.A.D.C., Maya, D.M.Y., Lora, E.E.S., 2023. Global bioenergy potentials projections for 2050. Biomass Bioenergy 09619534. 170, 106721. doi:10.1016/j.biombioe.2023.106721.

Grangeia, C., Santos, L., Lazaro, L.L.B., 2021. The Brazilian biofuel policy (RenovaBio) and its uncertainties: an assessment of technical, socioeconomic and institutional aspects. Energy Convers. Manag. X 13, 100156. doi:10.1016/j.ecmx.2021.100156.

Hameed, Z., Aslam, M., Khan, Z., Maqsood, K., Atabani, A., Ghauri, M., Khurram, M.S., Rehan, M., Nizami, A., 2020. Gasification of municipal solid waste blends with biomass for energy production and resources recovery: current status, hybrid technologies and innovative prospects. Renew. Sustain. Energy Rev. 136, 110375. doi:10.1016/j.rser.2020.110375.

Hassan, S.S., Williams, G.A., Jaiswal, A.K., 2018. Lignocellulosic biorefineries in Europe: current state and prospects. Trends Biotechnol. 37 (3), 231–234. doi:10.1016/j.tibtech.2018.07.002.

Hoang, A.T., Pandey, A., Huang, Z., Nižetić, S., Le, A.T., Nguyen, X.P., 2022. Biofuels an option for agro-waste management. In: Rehman, H.K., Bandh, A., Malla, S.A., Mohammad, F., Mehmood, A. (Eds.), Environmental Sustainability of Biofuels: Prospects and Challenges. Elsevier, Viet Nam, pp. 27–47. Available from: https://www.sciencedirect.com/book/9780323911597. doi:10.1016/B978-0-323-91159-7.00011-4.

International Energy Agency (IEA), 2023. *National Policy on Biofuels (2022 Amendment)*. https://www.iea.org/policies/17006-national-policy-on-biofuels-2022-amendment.

Jayed, M., Masjuki, H., Kalam, M., Mahlia, T., Husnawan, M., Liaquat, A., 2010. Prospects of dedicated biodiesel engine vehicles in Malaysia and Indonesia. Renew. Sustain. Energy Rev. 15 (1), 220–235. doi:10.1016/j.rser.2010.09.002.

Khan, A., Niazi, M.B.K., Ansar, R., Jahan, Z., Javaid, F., Ahmad, R., Anjum, H., Ibrahim, M., Bokhari, A., 2023. Thermochemical conversion of agricultural waste to hydrogen, methane, and biofuels: a review. Fuel. 00162361. 351. Available from: https://www.journals.elsevier.com/fuel/. doi:10.1016/j.fuel.2023.128947.

Kulisic, B., Gagnon, B., Schweinle, J., Van Holsbeeck, S., Brown, M., Simurina, J., Dimitriou, I., McDonald, H., 2021. The contributions of biomass supply for bioenergy in the post-COVID-19 recovery. Energies. 1996-1073. 14 (24), 8415. doi:10.3390/en14248415.

Kumar, J.A., Sathish, S., Prabu, D., Renita, A.A., Saravanan, A., Deivayanai, V.C., Anish, M., Jayaprabakar, J., Baigenzhenov, O., Hosseini-Bandegharaei, A., 2023. Agricultural waste biomass for sustainable bioenergy production: feedstock, characterization and pretreatment methodologies. Chemosphere. 18791298. 331. Available from: https://www.elsevier.com/locate/chemosphere. doi:10.1016/j.chemosphere.2023.138680.

Low, Y.W., Yee, K.F., 2021. A review on lignocellulosic biomass waste into biochar-derived catalyst: current conversion techniques, sustainable applications and challenges. Biomass Bioenergy. 18732909. 154. Available from: https://www.journals.elsevier.com/biomass-and-bioenergy/. doi:10.1016/j.biombioe.2021.106245.

Masjuki, H.H., Kalam, M.A., Mofijur, M., Shahabuddin, M., 2013. Biofuel: policy, standardization and recommendation for sustainable future energy supply. Energy Procedia. 18766102. 42, 577–586. doi:10.1016/j.egypro.2013.11.059.

Mehmood, U., 2021. Contribution of renewable energy towards environmental quality: the role of education to achieve sustainable development goals in G11 countries. Renew. Energy. 09601481. 178, 600–607. doi:10.1016/j.renene.2021.06.118.

Mujtaba, M., Fraceto, L.F., Fazeli, M., Mukherjee, S., Savassa, S.M., de Medeiros, G.A., Do Espírito Santo Pereira, A., Mancini, S.D., Lipponen, J., Vilaplana, F., 2023. Lignocellulosic biomass from agricultural waste to the circular economy: a review with focus on biofuels, biocomposites and bioplastics. J. Clean. Prod. 09596526. 402. Available from: https://www.journals.elsevier.com/journal-of-cleaner-production. doi:10.1016/j.jclepro.2023.136815.

Ogunkunle, O., Ahmed, N.A., 2019. A review of global current scenario of biodiesel adoption and combustion in vehicular diesel engines. Energy Rep. 23524847. 5, 1560–1579. Available from: https://www.journals.elsevier.com/energy-reports/. doi:10.1016/j.egyr.2019.10.028.

Oyama, A., Masutani, S., 2017. A review of the methane hydrate program in Japan. Energies. 1996-1073. 10 (10), 1447. doi:10.3390/en10101447.

Pearson, R.J., Turner, J.W.G., 2014. Improving the Use of Liquid Biofuels in Internal Combustion Engines. Elsevier BV, pp. 389–440. doi:10.1533/9780857097385.2.389.

Pravin, R., Baskar, G., 2024. Technoeconomic and carbon footprint analysis of simulated industrial scale biodiesel production process from mixed macroalgal and non-edible seed oil using sulphonated zinc doped recyclable biochar catalyst. Bioresour. Technol. 09608524. 395, 130351. doi:10.1016/j.biortech.2024.130351.

Ravichandran, P., Rajendran, N., Al-Ghanim, K.A., Govindarajan, M., Gurunathan, B., 2023. Investigations on evaluation of marine macroalgae *Dictyota bartayresiana* oil for industrial scale production of biodiesel through technoeconomic analysis. Bioresour. Technol. 09608524. 374, 128769. doi:10.1016/j.biortech.2023.128769.

Rupesh, S., Deepanraj, B., Rag, R.L., 2024. Valorization of lignocellulosic biomass through biorefinery concepts. In: Waste Valorization for Bioenergy and Bioproducts: Biofuels, Biogas, and Value-Added Products. Elsevier, India, pp. 461–503. Available from: https://www.sciencedirect.com/book/9780443191718. doi:10.1016/B978-0-443-19171-8.00015-8.

Sangeetha, B., Pravin, R., Baskar, G., 2024. Technoeconomic and environmental impact assessment of industrial scale biodiesel production process from algal oil using a reusable basic ionic liquid magnetic nanocatalyst. Energy Convers. Manag. 01968904. 314, 118668. doi:10.1016/j.enconman.2024.118668.

Saravanan, A., Kumar, P.S., Jeevanantham, S., Karishma, S., Vo, D.N., 2021. Recent advances and sustainable development of biofuels production from lignocellulosic biomass. Bioresour. Technol. 344, 126203. doi:10.1016/j.biortech.2021.126203.

Saravanan, A.P., Pugazhendhi, A., Mathimani, T., 2020. A comprehensive assessment of biofuel policies in the BRICS nations: implementation, blending target and gaps. Fuel 272, 117635. doi:10.1016/j.fuel.2020.117635.

Shabbir, M., Anwar, M.N., Saif, H., Tahir, E., Tahir, A., Rehan, M., Tanveer, R., Aghbashlo, M., Tabatabaei, M., Nizami, A.S., 2023. Policy and regulatory constraints in the biodiesel production and commercialization. In: Tabatabaei, M., Nizami, A.S. (Eds.), Sustainable Biodiesel: Real-World Designs, Economics, and Applications. Elsevier, Pakistan, pp. 357–372. Available from: https://shop.elsevier.com/books/sustainable-biodiesel/tabatabaei/978-0-12-820361-3. doi:10.1016/B978-0-12-820361-3.00007-3.

Sharma, A., Srikavya, B.P., Urade, A., Joshi, A., Narain, R.S., Dwarakanath, V., Alam, T., Prasad, R.S., 2023a. Economic and environmental impacts of biofuels in Indian context. In: *Materials Today Proceedings*. doi:10.1016/j.matpr.2023.05.118.

Sharma, P., Bano, A., Singh, S.P., Tong, Y.W., 2023b. Lignocellulosic biomass production: biodiversity benefits and threats. In: Singh, K., Milton, C.R., Ozgul, C., Biodiversity and Bioeconomy: Status Quo, Challenges, and Opportunities. Elsevier, Singapore, pp. 425–443. Available from: https://www.sciencedirect.com/book/9780323954822. doi:10.1016/B978-0-323-95482-2.00019-5.

Sidana, A., Yadav, S.K., 2022. Recent developments in lignocellulosic biomass pretreatment with a focus on eco-friendly, non-conventional methods. J. Clean. Prod. 09596526. 335, 130286. doi:10.1016/j.jclepro.2021.130286.

Song, J., Liu, C., Xing, J., Yang, W., Ren, J., 2022. Linking bioenergy production by agricultural residues to sustainable development goals: prospects by 2030 in China. Energy Convers. Manag. 276, 116568. doi:10.1016/j.enconman.2022.116568.

Su, Y., Zhang, P., Su, Y., 2015. An overview of biofuels policies and industrialization in the major biofuel producing countries. Renew. Sustain. Energy Rev. 13640321. 50, 991–1003. doi:10.1016/j.rser.2015.04.032.

Usmani, R.A., Mohammad, A.S., Ansari, S.S., 2023. Comprehensive biofuel policy analysis framework: a novel approach evaluating the policy influences. Renew. Sustain. Energy Rev. 18790690. 183. Available from: https://www.journals.elsevier.com/renewable-and-sustainable-energy-reviews. doi:10.1016/j.rser.2023.113403.

Wolinetz, M., Hein, M., 2017. Biofuels in Canada 2017: Tracking biofuel consumption, feedstocks and avoided greenhouse gas emissions. Vancouver BC V7X 1A1, Navius Research Inc.

Zhao, J., 2015. Development of China's biofuel industry and policy making in comparison with international practices. Sci. Bull. 20959273. 60 (11), 1049–1054. doi:10.1007/s11434-015-0803-2.

CHAPTER 9

Future prospects and challenges for bioenergy from agricultural residues

S. Prasanna Raj Yadav[a], K. Thavasilingam[a], B. Bharathiraja[b], and R. Praveen kumar[c]
[a]Department of Mechanical Engineering, Easwari Engineering College, Ramapuram, Chennai, Tamil Nadu, India
[b]Department of Biotechnology, Easwari Engineering College, Chennai, Tamil Nadu, India
[c]Department of Biotechnology, Arunai Engineering College, Chennai, Tamil Nadu, India

9.1 Current state and future prospects of bioenergy from agricultural residues

Agricultural residues are now at the forefront of renewable bioenergy sources as a result of the shift to sustainable energy systems. Crop residues, animal manure, and organic municipal trash are examples of agricultural biomass, which is a readily available but underutilized feedstock for decentralized energy production. Countries all across the world have begun to evaluate and utilize this potential, but the size of use is still disproportionately small when compared to the availability of resources. For example, the technological bioenergy potential in Russia is estimated to be around 2225.4 PJ, with crop wastes accounting for 42% of this potential. However, only 12% of this potential is currently being utilized. Despite having plenty of resources, this only accounts for 30% of Russia's overall heat and electricity use, suggesting a serious underutilization (Namsaraev et al., 2018).

In a similar vein, China has significantly improved its use of GIS-based spatial studies to measure agricultural residue availability. With a corresponding bioenergy potential of 7.4 EJ yearly, the net available crop residue is predicted to be 505.5 million tonnes annually, taking into account transportation costs and environmental constraints. As of 2009, this amounts to 8.27% of China's total energy consumption (Jiang et al., 2012). Using second-generation biomass, such as agricultural waste, as a substitute for food-based energy crops is becoming more and more popular on a global scale. By facilitating waste valorization and nutrient recycling, agricultural leftovers are in line with the concepts of the circular economy, in contrast

Bioenergy from Agricultural Residues
https://doi.org/10.1016/B978-0-443-24726-2.00010-9

to first-generation bioenergy sources that compete with food supply. Through the creation of jobs, their use aids in waste management, the mitigation of climate change, and rural development (Bonatto et al., 2020). Geographic variability, varying methodological approaches, and changing land-use practices make it difficult to estimate the true bioenergy potential of agricultural residues. While remote sensing and GIS-based models provide more precision by adding dynamic environmental variables, traditional statistical data-based methods frequently overestimate potential by ignoring conservation and harvesting limits (Long et al., 2013). Crop residue biomass is a key element of the renewable energy strategy in India, where agriculture is a major economic driver. An estimated 686 million tonnes of gross crop residue are produced in India each year, of which 234 million tonnes are surplus and technically suitable for energy production. Notably, rice and sugarcane account for the largest portion of surplus residues, and availability varies significantly between states (Hiloidhari et al., 2014). This potential is further expanded by recent assessments of biomass leftovers, such as urban organic waste and livestock manure (Deep Singh et al., 2022). Despite being a viable and sustainable feedstock for the generation of bioenergy, agricultural leftovers are now only used in a restricted number of places because of legislative, infrastructure, and technological obstacles. Future prospects rely on better mapping of biomass resources, more effective technological interventions, and more robust policy frameworks that allow bioenergy's contribution to low-carbon energy transitions to be fully realized. Nations such as China, India, and Russia serve as excellent examples of the opportunities and difficulties that lie ahead, and their experiences offer important lessons for global energy planning that is based on the value-adding of agricultural waste.

China and India, two nations with sizable agricultural sectors, show significant promise for producing bioenergy from their agricultural waste streams, as shown in Fig. 9.1. China is thought to have a substantial amount of bioenergy potential from residues, possibly surpassing 2000 PJ/year, due to its extensive agricultural acreage and varied crop production. Similar to this, India, a significant agricultural economy, shows a significant amount of capability; according to certain assessments, it could produce >1300 PJ of surplus crop residue annually. Brazil, which is known for having a thriving bioethanol sector, also demonstrates a tremendous potential for bioenergy from agricultural residues, highlighting its position as a leader in renewable energy. The worldwide agricultural residue bioenergy landscape is significantly influenced by important European nations such as Germany

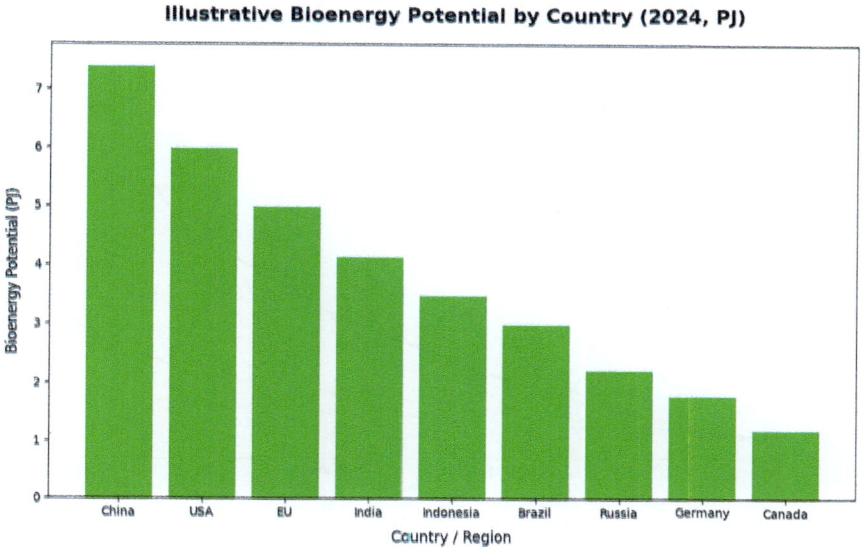

Fig. 9.1 Global map of agricultural residue-based bioenergy projects (2024).

and France as well as other countries with various agricultural practices, such as the United States.

From Errera et al. (2023), Fig. 9.2 shows estimates of the world's bioenergy supply from agricultural residues under three different future scenarios. The predicted potential, which reflects current trajectory trends, rises slightly to 64 EJ by 2050 under the business as usual scenario. Agri-waste recovery and processing have improved moderately under the optimistic trends scenario, reaching 125 EJ/year. The upper bound—313 EJ/year—made possible by developments in biotechnology, residue collection technologies, and policy frameworks that support biomass-based energy systems is demonstrated in the full adaptation response (FAR) scenario. In addition, according to the FAR pathway, bioenergy may substantially reduce reliance on fossil fuels by meeting >37% of the world's total primary energy consumption. It is crucial to remember that these estimates are dependent on the adoption of strong sustainability frameworks, preventing adverse effects on water resources, food security, and biodiversity. To achieve this anticipated potential, further research and development in feedstock management, conversion efficiency, and end-use applications will be essential.

Based on data provided from the FAO and IPCC, Table 9.1 analyses the energy yields and residue-to-product ratios of the main residues, which include rice straw, wheat straw, sugarcane bagasse, and maize stalk.

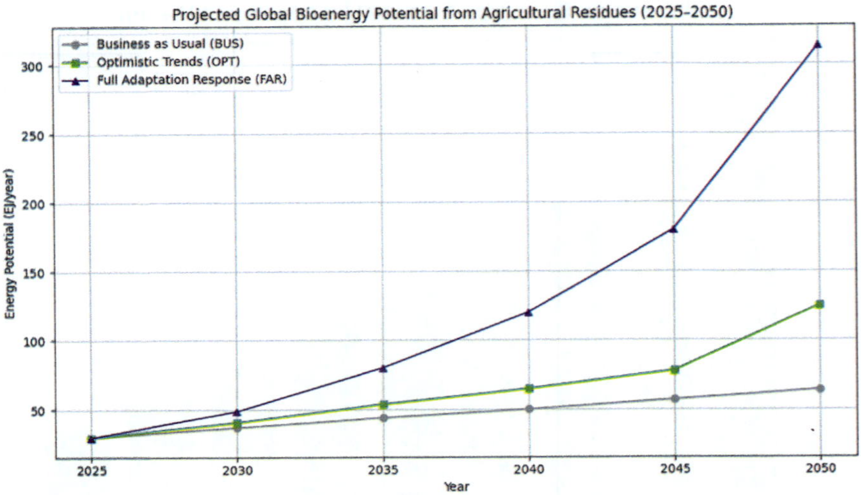

Fig. 9.2 Projected growth of bioenergy from agri-residues (2025–2050).

Table 9.1 Comparative overview of major agricultural residues and their energy yields.

Residue	Residue-to-product ratio	Energy content (MJ/kg)
Rice straw	0.87	~15
Wheat straw	0.84	~14
Sugarcane bagasse	0.65	~12
Maize stalk	0.82	~14

Utilization is still low despite the enormous potential: When total output, sustainable availability, and actual usage are compared in Table 9.2, it becomes clear that, despite residues making up a sizable portion of their energy mix, India only uses around 17% of its residue potential, China 8.3%, and Russia 12%.

9.2 Technological advances and challenges

Innovations in biomass conversion technology and the solving of significant practical issues are necessary for the shift from fossil fuels to bioenergy. Current bioenergy conversion techniques, including thermochemical, biochemical, and innovative hybrid processes, have advanced considerably

Table 9.2 Regional availability vs. utilization of agri-biomass for energy (2023 data).

Region/country	Total residue production (EJ/year)	Sustainable availability (%)	Utilized for energy (%)
India	~4.15	~50	~17
China	~7.4	~60	~8.3
Russia	~2.225	~30	~12
EU (aggregate)	~5.0	~40	~20

but are still mostly limited to laboratory or pilot scales, as Banerjee (2023) pointed out. New methods for increasing conversion efficiency and waste valorization include biochemical looping combustion, and both photo- and dark fermentation. These techniques can be improved even more by combining machine learning (ML) and artificial intelligence (AI) to optimize system design and processing parameters, increasing the predictability and adaptability of bioenergy production.

According to Shahzad et al. (2024), agricultural waste that is high in cellulose, hemicellulose, and lignin can be effectively transformed through thermochemical (such as pyrolysis and gasification) and biochemical (such as anaerobic digestion and fermentation) processes into a variety of bioenergy forms, including bio-oil, biogas, and biochar. However, each strategy has unique drawbacks. For example, thermochemical techniques necessitate expensive equipment and significant energy inputs, whereas biochemical methods are limited by lengthy processing periods and low yields. To solve these inefficiencies and scale the technologies, there is an urgent need for research in thermal integration, feedstock pretreatment, and catalyst development. Singh et al. (2025) provide a broader perspective on the socio-environmental implications of bioenergy, reiterating that biomass-based fuels can significantly reduce greenhouse gas emissions and improve rural livelihoods.

Key issues identified by their investigation include water use, regulatory obstacles, and conflicts between food and fuel. It is also explored how cutting-edge technologies, such as hydrogen fuel cells and battery electric integration, might enhance the infrastructure for biofuels. The industry could undergo radical change if machine intelligence is used for process control, lifespan assessment, and policy modeling. Examining the particular instance of agricultural residue conversion, Awogbemi et al. (2024) promote strict collecting, sorting, and pretreatment procedures as a requirement for

Fig. 9.3 Bioenergy conversion technologies for agricultural residues.

producing bioenergy on a large scale. The viability of sustainable waste-to-energy pathways is demonstrated by technologies that produce bioethanol, biogas, biohydrogen, and biojet fuel from agricultural waste streams. To guarantee consistency and quality in bioenergy production, it urges additional interdisciplinary research that integrates sensing, automation, and process control, as shown in Fig. 9.3.

9.3 Policy and economic considerations for commercialization

The convergence of institutional support mechanisms, economic incentives, and policy frameworks has a significant impact on the commercialization of bioenergy derived from agricultural leftovers. Despite the maturity of technological feasibility, economic and policy-related obstacles sometimes impede the scaling of these systems. The lack of coordinated policies that connect agricultural waste management to energy production, carbon reduction, and improving rural livelihoods has caused fragmentation across implementation channels in many emerging nations. Nonetheless, a slow transition is being noted in the direction of a more comprehensive policy ecosystem meant to promote the use of bioenergy as depicted in Fig. 9.4.

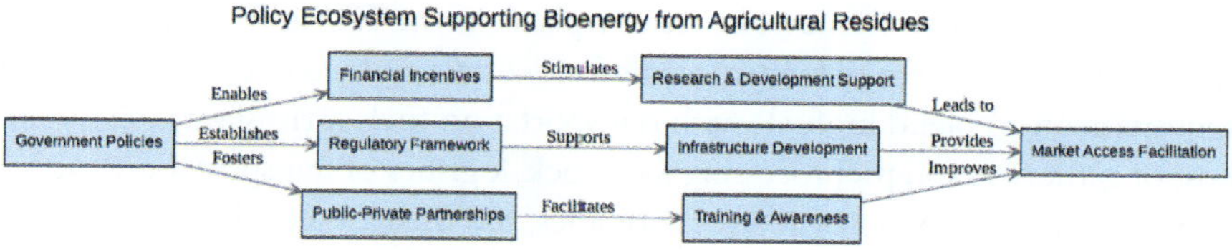

Fig. 9.4 Policy ecosystem supporting bioenergy from agricultural residues.

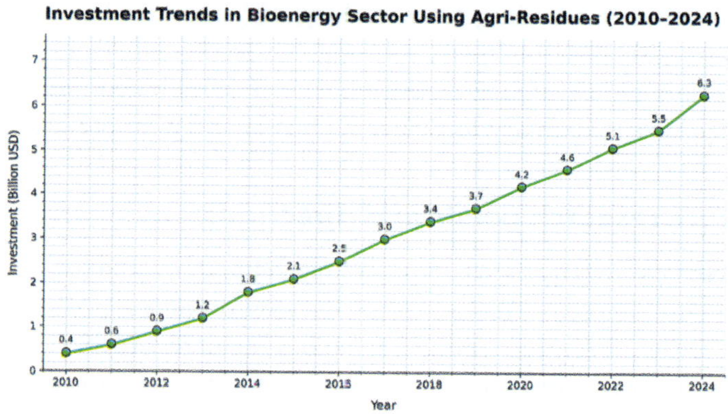

Fig. 9.5 Investment trends in the bioenergy sector using agri-residues (2010–2024).

Circular bioeconomy ideas, which use agricultural leftovers as dual-purpose feedstocks for energy and environmental sustainability, are being adopted by governments more and more.

For example, in India, decentralized biogas production units using agri-residues have been encouraged by policies such as the National Bio-Energy Mission andSustainable Alternative Towards Affordable Transportation scheme (Kapoor et al., 2020). The practical integration of these projects is frequently lacking due to insufficient investment pipelines and policy execution at the state or regional levels, despite the fact that they are supported by climate action targets and rising energy demand (Vasco-Correa et al., 2018). Over the past 10 years, investment in this industry has increased, with large upticks during times of targeted fiscal support or policy certainty, as shown in Fig. 9.5. However, there are still major obstacles to commercialization, including return on investment risks, land use conflicts, and volatility in biomass prices (Pingali and Rosegrant, 1995).

Similar legislative frameworks are in place in the EU and Southeast Asia, specifically in the form of feed-in tariffs for energy produced from biomass and requirements to blend biofuels with transportation fuels (Srivastava, 2024). The cost of gathering and preprocessing feedstock, logistics of transportation, capital expenditures for bioconversion technologies, and the market value of the energy product (biogas, bio-oil, electricity, etc.) are all interrelated factors that affect the economic feasibility of agri-residue-based bioenergy. According to studies, areas with inadequate infrastructure for biomass aggregation or irregular government subsidies have a hard time reaching the economic break-even point (Singh et al., 2023). A multistakeholder policy ecosystem that incorporates land-use planning, agricultural productivity, rural development, and renewable energy mandates is necessary to reduce such risks and promote scale-up.

9.4 Future research needs and directions

Future research initiatives must be strategically aligned with technology advancements, legislative frameworks, and environmental goals due to the dynamic nature of biomass energy research. The increasing pace of research in biomass energy worldwide is demonstrated by a bibliometric analysis of >33,000 articles published between 1998 and 2013, with a focus on the transition from conventional physical conversion techniques to more sophisticated thermochemical and biochemical approaches. Growing interest in emerging subjects such as biohydrogen, biodiesel, and microbial fuel cells is exposing developing research hotspots that represent both technological maturity and socio-environmental significance. With organizations such as the University of Illinois and the Chinese Academy of Sciences playing crucial collaborative roles, the United States, China, and India dominate the world in academic production, demonstrating a globally interconnected research ecosystem (Mao et al., 2015).

Up until 2050, bioenergy is expected to continue to be a key component of Europe's renewable energy policy. If local biomass resources are used to their full capacity, estimates indicate that energy import dependency may be reduced by 22%. The economic feasibility of feedstocks, technological adaptability, and changes in energy policy are some of the major uncertainties that call for further research into supply-demand dynamics and global biomass trade. In order to address these complex uncertainties, future research must use integrative and harmonized modeling tools that evaluate both socioeconomic viability and technical potential (Mandley et al., 2020). In addition, focused content analysis in the bioenergy space identifies important research areas related to biogas, anaerobic digestion, and biomass valorization associated with agriculture. However, there are still gaps in our understanding, especially when it comes to investigating underutilized agricultural leftovers and cutting-edge technology such as microbial fuel cells. These gaps represent a strong case for multidisciplinary research that combines environmental engineering, biotechnology, and agricultural sciences (Kaur et al., 2025). There are practical difficulties as well as technical possibilities in turning agro-farm waste into bioenergy. Although well-established conversion technologies such as transesterification and thermochemical procedures have drawbacks in terms of cost, efficiency, and scalability. The goal of future research must be to overcome these constraints by using hybrid processing methods, sophisticated material science, and reactor design. Implementation potential will be improved by a comparative analysis of these techniques under various feedstock types and geographical resource availability (Asfaw et al., 2024). Importantly, a revolutionary change has occurred with the incorporation of automation, AI, and ML into biomass transformation systems. Particularly in anaerobic digestion and bioethanol production, these technologies allow for real-time monitoring, intelligent process control, and predictive analytics. But there are important issues that need to be resolved, such as infrastructure development, data standardization, and the requirement for multidisciplinary workforce training. In order to fully reap the benefits of AI/ML in biomass conversion, future research should prioritize the development of data-driven reactor management tools, automated experimental platforms, and intelligent decision support systems (Ananda et al., 2025). In summary, a systems-level approach that integrates domain expertise, digital innovation, and policy alignment is essential for the future of biomass energy research.

9.5 Conclusion

The thorough investigation of bioenergy produced from agricultural residues highlights its enormous potential to address the problems of environmental sustainability, rural development, and agricultural waste management while also making a significant contribution to the global renewable energy portfolio. The efficiency, limitations, and deployment maturity levels of the many technologies used in the conversion of agricultural residues into bioenergy, including thermochemical and biochemical routes, have been described in detail in the manuscript. It is clear from the comparative analysis of different biomass sources, such as maize stover, rice husks, wheat straw, and sugarcane bagasse, that when developing localized bioenergy strategies, regional differences in feedstock availability and energy yield must be taken into account.

Furthermore, improvements in integrated biorefineries, reactor design, and pretreatment procedures keep enhancing the energy recovery and environmental impact of bioenergy systems. Significant technological and financial obstacles still exist despite the positive developments, especially in the areas of feedstock heterogeneity, biomass collecting logistics, and the scalability of sophisticated conversion procedures. Furthermore, the widespread commercialization of these technologies is hampered by the absence of standardized rules, shifting market dynamics, and a lack of adequate financial incentives. The transition of pilot-scale projects to commercial viability is facilitated by targeted subsidies and supportive policy frameworks, as this study highlights. To improve resilience and adaptation across various agro-climatic zones, future research must concentrate on decentralized energy models, AI-integrated process optimization, and hybrid bioenergy systems.

The anticipated growth trends show that agricultural residue-based bioenergy can provide a low-carbon substitute for fossil fuels and make a substantial contribution to supplying the world's energy needs by 2050, especially under optimistic and adaptation-driven scenarios. But achieving this potential calls for a comprehensive strategy that blends political dedication, socioeconomic inclusiveness, and scientific innovation. In order to fully utilize agricultural residues and turn them from a waste burden into a sustainable energy resource—a crucial step in the shift to a circular and carbon-neutral bioeconomy—this manuscript emphasizes the necessity of multidisciplinary and multisectoral collaboration.

References

Ananda, A., Sujeeth, R.K., Archana, S., 2025. Integrating automation in biomass transformation: opportunities, challenges, and future directions. Bioenerg. Res. 1939-1242. 18 (1). doi:10.1007/s12155-025-10864-6.

Asfaw, B.T., Gari, M.T., Jeyapaul, A.S., Jayakumar, M., Baskar, G., 2024. Volarization of agro farm yard waste into bioenergy: present state and future directions. In: Biofuels and Bioenergy: Biorefinery and Circular Bioeconomy Approaches. Elsevier, Ethiopia, 173–201. Available from https://www.sciencedirect.com/book/9780443292545. doi:10.1016/B978-0-443-29254-5.00008-4.

Awogbemi, O., Kallon, D.V.V., Ray, R.C., 2024. Conversion of Agriculture Residues for Bioenergy Production. Springer Science and Business Media LLC, pp. 273–294. doi:10.1007/978-3-031-61133-9_12.

Banerjee, N., 2023. Biomass to energy—an analysis of current technologies, prospects, and challenges. Bioenerg. Res. 19391242. 16 (2), 683–716. Available from: https://www.springer.com/journal/12155. doi:10.1007/s12155-022-10500-7.

Bonatto, C., Camargo, A.F., Scapini, T., Stefanski, F.S., Alves, S.L., Müller, C., Fongaro, G., Treichel, H., 2020. Biomass to bioenergy research: current and future trends for biofuels. In: Recent Developments in Bioenergy Research. Elsevier, Brazil, 1–17. Available from https://www.elsevier.com/books/recent-developments-in-bioenergy-research/gupta/978-0-12-819597-0. doi:10.1016/B978-0-12-819597-0.00001-5.

Deep Singh, A., Gajera, B., Sarma, A.K. 2022. Appraising the availability of biomass residues in India and their bioenergy potential. Waste Manag. 0956053X. 152, 38–47. doi:10.1016/j.wasman.2022.08.001.

Errera, M.R., Dias, T.A.D.C., Maya, D.M.Y., Lora, E.E.S., 2023. Global bioenergy potentials projections for 2050. Biomass Bioenerg. 09619534. 170, 106721. doi:10.1016/j.biombioe.2023.106721.

Hiloidhari, M., Das, D., Baruah, D.C., 2014. Bioenergy potential from crop residue biomass in India. Renew. Sustain. Energy Rev. 13640321. 32, 504–512. doi:10.1016/j.rser.2014.01.025.

Jiang, D., Zhuang, D., Jinying, F., Huang, Y., Wen, K., 2012. Bioenergy potential from crop residues in China: availability and distribution. Renew. Sustain. Energy Rev. 13640321. 16 (3), 1377–1382. doi:10.1016/j.rser.2011.12.012.

Kapoor, R., Ghosh, P., Kumar, M., Sengupta, S., Gupta, A., Kumar, S.S., Vijay, V., Kumar, V., Vijay, V.K., Pant, D., 2020. Valorization of agricultural waste for biogas based circular economy in India: a research outlook. Bioresour. Technol. 18732976. 304. Available from: https://www.elsevier.com/locate/biortech. doi:10.1016/j.biortech.2020.123036.

Kaur, S., Kumar, R., Singh, K., 2025. Agriculture waste and bioenergy: a content analysis review and visualization. In: Fourth International Conference on Power, Control and Computing Technologies, ICPC2T 2025. Institute of Electrical and Electronics Engineers Inc., India, pp. 257–262. http://ieeexplore.ieee.org/xpl/mostRecentIssue.jsp?punumber=10958571. doi:10.1109/ICPC2T63847.2025.10958710.

Long, H., Xiaobing, L., Wang, H., Jia, J., 2013. Biomass resources and their bioenergy potential estimation: a review. Renew. Sustain. Energy Rev. 13640321. 26, 344–352. doi:10.1016/j.rser.2013.05.035.

Mandley, S.J., Daioglou, V., Junginger, H.M., van Vuuren, D.P., Wicke, B., 2020. EU bioenergy development to 2050. Renew. Sustain. Energy Rev. 13640321. 127 (109858). doi:10.1016/j.rser.2020.109858.

Mao, G., Zou, H., Chen, G., Huibin, D., Zuo, J., 2015. Past, current and future of biomass energy research: a bibliometric analysis. Renew. Sustain. Energy Rev. 13640321. 52, 1823–1833. doi:10.1016/j.rser.2015.07.141.

Namsaraev, Z.B., Gotovtsev, P.M., Komova, A.V., Vasilov, R.G., 2018. Current status and potential of bioenergy in the Russian Federation. Renew. Sustain. Energy Rev. 13640321. 81, 625–634. doi:10.1016/j.rser.2017.08.045.

Pingali, P.L., Rosegrant, M.W., 1995. Agricultural commercialization and diversification: processes and policies. Food Policy. 03069192. 20 (3), 171–185. doi:10.1016/0306-9192(95)00012-4.

Shahzad, H.M.A., Asim, Z., Khan, S.J., Almomani, F., Mahmoud, K.A., Mustafa, M.R.U., Rasool, K., 2024. Thermochemical and biochemical conversion of agricultural waste for bioenergy production: an updated review. Discov. Environ. 27319431. 2 (1). Available from: https://link.springer.com/journal/44274. doi:10.1007/s44274-024-00171-w.

Singh, R., Gaur, A., Soni, P., Jain, R., Pant, G., Kumar, D., Kumar, G., Shamshuddin, S.Z.M., Mubarak, N.M., Dehghani, M.H., Suhas, K.A., 2025. A review of biofuels and bioenergy production as a sustainable alternative: opportunities, challenges and future perspectives. J. Environ. Health Sci. Eng. 2052336X. 23 (2). Available from: https://www.springer.com/journal/40201. doi:10.1007/s40201-025-00946-0.

Singh, S., Pant, K.K., Krishania, M., 2023. Current perspective for bio-oil production from agricultural residues in commercialization aspect: a review. J. Anal. Appl. Pyrolysis. 01652370. 175, 106160. doi:10.1016/j.jaap.2023.106160.

Srivastava, P., 2024. Government Initiative and Policy for Agricultural Waste Utilization as Biofuel. Springer Science and Business Media LLC, pp. 273–304. doi:10.1007/978-981-99-8244-8_13.

Vasco-Correa, J., Khanal, S., Manandhar, A., Shah, A., 2018. Anaerobic digestion for bioenergy production: global status, environmental and techno-economic implications, and government policies. Bioresour. Technol. 18732976. 247, 1015–1026. Available from: https://www.elsevier.com/locate/biortech. doi:10.1016/j.biortech.2017.09.004.

Index

Note: Page numbers followed by "*f*" indicate figures, "*t*" indicate tables.

H

I

J

L

M

N

O

P

R

S

T

Printed and bound by CPI Group (UK) Ltd, Croydon, CR0 4YY
25/07/2026
02176694-0001